AF905660

# essentials

## MyCopy

Dear MyCopy Customer,

This Springer Nature book is a high quality monochrome, digitally printed version of the eBook, which is accessible to you on SpringerLink. As a result of your library investing in licensing at least one Springer Nature eBook subject collection, this copy is provided to you at a highly discounted price from the usual softcover edition.

MyCopy books are only offered to patrons of a library with access to at least one Springer Nature eBook subject collection and are strictly for individual use only. Personal resale or library shelving of the MyCopy book is in breach of this agreement.

You may cite this book by referencing the bibliographic data and/or the DOI (Digital Object Identifier) found in the front-matter. This book is an exact but monochrome copy of the print version of the eBook on SpringerLink.
www.springernature.com/mycopy

*essentials* liefern aktuelles Wissen in konzentrierter Form. Die Essenz dessen, worauf es als „State-of-the-Art" in der gegenwärtigen Fachdiskussion oder in der Praxis ankommt. *essentials* informieren schnell, unkompliziert und verständlich

- als Einführung in ein aktuelles Thema aus Ihrem Fachgebiet
- als Einstieg in ein für Sie noch unbekanntes Themenfeld
- als Einblick, um zum Thema mitreden zu können

Die Bücher in elektronischer und gedruckter Form bringen das Fachwissen von Springerautor*innen kompakt zur Darstellung. Sie sind besonders für die Nutzung als eBook auf Tablet-PCs, eBook-Readern und Smartphones geeignet. *essentials* sind Wissensbausteine aus den Wirtschafts-, Sozial- und Geisteswissenschaften, aus Technik und Naturwissenschaften sowie aus Medizin, Psychologie und Gesundheitsberufen. Von renommierten Autor*innen aller Springer-Verlagsmarken.

Patric U. B. Vogel

# COVID-19: Search for a vaccine

Springer

Patric U. B. Vogel
Vogel Pharmopex24
Cuxhaven, Germany

ISSN 2197-6708 ISSN 2197-6716 (electronic)
essentials

https://doi.org/10.1007/978-3-658-38931-4

© The Editor(s) (if applicable) and The Author(s), under exclusive licence to Springer Fachmedien Wiesbaden GmbH, part of Springer Nature 2023
This book is a translation of the original German edition „COVID-19: Suche nach einem Impfstoff " by Vogel, Patric U. B., published by Springer Fachmedien Wiesbaden GmbH in 2021. The translation was done with the help of artificial intelligence (machine translation by the service DeepL.com). A subsequent human revision was done primarily in terms of content, so that the book will read stylistically differently from a conventional translation. Springer Nature works continuously to further the development of tools for the production of books and on the related technologies to support the authors.
This work is subject to copyright. All rights are solely and exclusively licensed by the Publisher, whether the whole or part of the material is concerned, specifically the rights of translation, reprinting, reuse of illustrations, recitation, broadcasting, reproduction on microfilms or in any other physical way, and transmission or information storage and retrieval, electronic adaptation, computer software, or by similar or dissimilar methodology now known or hereafter developed.
The use of general descriptive names, registered names, trademarks, service marks, etc. in this publication does not imply, even in the absence of a specific statement, that such names are exempt from the relevant protective laws and regulations and therefore free for general use.
The publisher, the authors, and the editors are safe to assume that the advice and information in this book are believed to be true and accurate at the date of publication. Neither the publisher nor the authors or the editors give a warranty, expressed or implied, with respect to the material contained herein or for any errors or omissions that may have been made. The publisher remains neutral with regard to jurisdictional claims in published maps and institutional affiliations.

This Springer Spektrum imprint is published by the registered company Springer Fachmedien Wiesbaden GmbH, part of Springer Nature.
The registered company address is: Abraham-Lincoln-Str. 46, 65189 Wiesbaden, Germany

# What You Can Find in this *essential*

- An introduction to the principle of old and new vaccine technologies
- The presentation of strengths and weaknesses of each technology
- An overview of the current vaccine projects against COVID-19 and the progress of vaccination with the first licensed vaccines
- The presentation of further aspects, from the willingness to vaccinate to the occurrence of mutations and an evaluation of previous vaccination side reactions
- An introduction to additional concepts such as sterilizing immunity, herd immunity, and immunity duration

# Contents

# About the Author

**Patric U. B. Vogel** is a biologist with many years of experience with various biopharmaceuticals, from live vaccines to DNA vaccines. He was a registered Qualified Person for several years and has regularly released coronavirus vaccines against animal diseases, among other things. Since 2019, he has been an author, trainer and consultant in the pharmaceutical field.

# 1 Introduction, Background and Properties of Coronaviruses

## 1.1 Background

**Vaccines** are one of the greatest achievements of modern medicine. What began in the eighteenth century with the fight against **smallpox** has experienced a phenomenal triumph in public health. It is estimated that the global use of vaccines prevents several million deaths each year, especially among children (CDC 2014). However, vaccines are not miracle cures that can easily and conveniently eradicate infectious diseases. To date, only two infectious diseases, smallpox and rinderpest, have been completely eliminated through intensive control measures, including vaccination campaigns (Hamilton et al. 2015). Many other infectious diseases, such as measles, are merely kept under control. Why, even after decades of vaccine use, are there not more infectious diseases that have been eradicated? For one thing, not all people, especially in developing countries, are vaccinated. This leaves a susceptible population within which pathogens "survive." Second, certain vaccinations do not confer **sterilizing immunity,** meaning that the vaccinated person is protected against disease but not against infection, which allows pathogens to spread. Some viruses are constantly changing through genetic processes, or jump from animals to humans from time to time. Although for these reasons complete eradication is illusory in many cases, vaccines help to control many infectious diseases.

The process of **vaccine development** – the development from idea to approval – is lengthy and takes on average more than 10 years. It follows that no vaccine is initially available against new infectious diseases. In the twentieth century, this risk was relatively manageable, often in the form of new **influenza pandemics** (pandemic: spread of a pathogen across different continents) with long time intervals.

© The Author(s), under exclusive license to Springer Fachmedien Wiesbaden GmbH, part of Springer Nature 2023

P. U. B. Vogel, *COVID-19: Search for a vaccine*, essentials,
https://doi.org/10.1007/978-3-658-38931-4_1

Increasing globalization, a growing earth population, climate changes, and the progressive invasion of wildlife habitats have fundamentally changed the probability of occurrence of infectious diseases.

Novel viruses or the return of known viruses are occurring at increasingly frequent intervals and are now more the rule than the exception. This is illustrated by the numerous events of the last decade, including, for example, swine flu, the discovery of MERS, the increase in dengue fever, the Ebola and Zika epidemics, and the spread of **COVID-19** since the end of 2019. Thus, humanity is constantly facing new threats and coronaviruses appear to be key players in this twenty-first century threat. New coronaviruses that appear in humans originate from animals. A disease that is transmitted from animals to other animals or humans is called a **zoonotic disease.** Coronaviruses have made the jump from animals to humans many times, with bats and rodents identified as the main source, sometimes with other animals as intermediate carriers (Corman et al. 2018).

The novel coronavirus **SARS-CoV-2** has posed unprecedented challenges and difficulties for health systems and the economy. Although similar measures had been taken a hundred years ago during the **Spanish flu** outbreak, global spread was slower and measures were limited to specific countries or regions (Vogel and Schaub 2021). The similar **SARS virus** in 2002/2003 was also combated and brought under control by massive infection control measures (Chan-Yeung and Xu 2003). The circumstances of the rapid spread of **COVID-19** required a rapid response. The first wave already caused considerable damage in some European countries. After calmer summer months, a severe second wave followed in autumn or winter 2020. The speed of the pathogen's spread also made it necessary in Germany to implement ever further extensions and tightenings from the first so-called **breakwater shutdown,** which was planned for only one month.

A central, if not the most important, component in the fight against this new disease **COVID-19** lies in the availability of **vaccines.** While at the beginning of the pandemic a successful vaccine development and approval within a year seemed illusory, at least a look at the animal kingdom gave a hope for a basic success, as there are several effective vaccines, e.g. against coronaviruses of poultry or pigs (OIE 2000; Gerdts and Zakhartchouk 2017). The goal of having a vaccine against COVID-19 as soon as possible has set in motion unprecedented activities by research groups and pharmaceutical companies. The status of these projects is summarized and continuously updated by the **World Health Organization (WHO)** (WHO 2020a). Although accelerating the process of development and approval by a factor of approximately 10 (1 year instead of 10 years) comes with risks, several **vaccine technologies** have celebrated phenomenal success. Currently, several vaccines in Europe and the rest of the world have already received one of several

**forms of approval** (e.g. emergency or conditional approval, from now on called licensure) and over 100 million people have been vaccinated worldwide. It should also be emphasized that high-risk groups such as the elderly can be very well protected against COVID-19 disease.

In this *essential,* we will learn about different types of vaccines, their advantages and disadvantages, and the projects currently underway to develop vaccines against **COVID-19.** In the first edition, in addition to presenting various technologies, risks were also discussed, among other things. This was intended as a cautionary warning, as there were various efforts to override the usual regularities in the first half of 2020. These included, for example, political efforts in the USA to achieve an approval date before the elections. The greatest danger at the time was that Phase III clinical trials would be skipped and mass vaccination would begin. With a few exceptions, where vaccines were already being used in certain occupational groups in some countries before preliminary phase III clinical results were available, common sense prevailed here, as the example of the European regulatory authority **EMA** has also shown. Currently, there is increasing evidence that the efficacy of **COVID-19 vaccines** is being confirmed in vaccination campaigns and that they have a safety profile comparable to proven vaccines. In this second edition, some of the vaccines are presented and further aspects such as the emergence of **viral mutations** are discussed.

## 1.2 Coronaviruses: History and Viral Properties

**Coronaviruses** were first described as the causative agent of an infectious disease in poultry in the 1930s. The disease was named **"infectious bronchitis"** (Bijlenga et al. 2004) and remains one of the most dangerous infectious diseases of poultry today. The first human coronaviruses were discovered in the 1960s (Kahn and McIntosh 2005). However, molecular biology analyses indicate that some coronaviruses were transmitted to humans several hundred years ago and have been circulating in our population ever since (Graham et al. 2013). Coronaviruses are very small, measuring 80–120 nm (Masters 2006). This is about a hundred times smaller than our body cells. For this reason, their morphology, or appearance, could only be described using a high-resolution technique called **electron microscopy**. Under the electron microscope, viruses have an oval shape with long projections on the surface (Fig. 1.1). This crown-like appearance was decisive for the naming of the coronaviruses (the Latin term corona means crown).

A total of seven human coronaviruses are known. Four of these coronaviruses (designated 229E, NL63, HKU1 and OC63) are found worldwide and cause typical

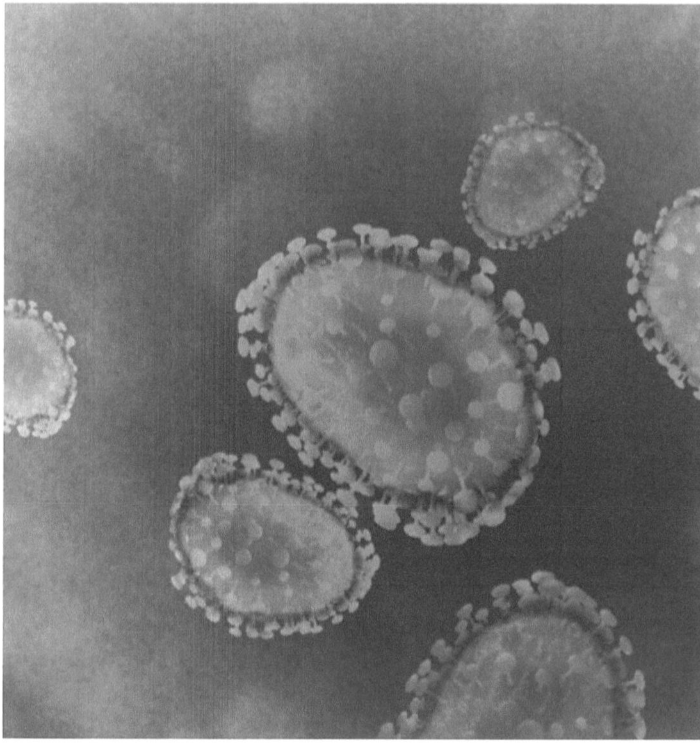

**Fig. 1.1** Electron microscopic image of coronaviruses. (With additionally inserted 3D effect; source: Adobe Stock, file no.: 329773404)

colds during the cold season, with which each of us has probably made unpleasant acquaintance one or more times in the course of our lives. These coronaviruses cause about 15% of all colds during the cold season (Kahn and McIntosh 2005; Greenberg 2016).

In addition, there are three other coronaviruses that were only transmitted from animals to humans or identified after the turn of the century. The diseases caused by them had or have a special medical significance because of the high mortality rate. The causative agent of **severe acute respiratory syndrome (SARS)**, first appeared in China in 2002. There were 8096 cases with 774 deaths (WHO 2004). SARS was declared to have been defeated in June 2003. Interestingly, SARS was not a one-time event of transmission from animals to humans. Diagnostics at the time were not as sophisticated as they are today, and it took several months to identify a coronavirus as the causative agent of this disease (Drosten et al. 2003). Subsequent analysis of serum samples (fluid obtained from blood samples that contains antibodies, among other things) showed that a year earlier, in 2001, healthy people in Hong Kong had antibodies to SARS viruses (Graham et al. 2013). These people must therefore have undergone an infection.

**Middle East Respiratory Syndrome (MERS)** was first described in 2012 (Zaki et al. 2012). Unlike SARS, there was no time-limited local outbreak. MERS is a constant threat because the virus circulates in dromedaries (one-humped camels), for which it is relatively harmless, and is always sporadically transmitted to humans. In total, there are now over 2500 known human cases of MERS (WHO 2020b). Based on retrospective analysis of serum samples, MERS viruses are thought to have been present in dromedaries since the early 1980s (de Wit et al. 2016). Antibodies to MERS viruses have been found in 70–100% of dromedaries in the Arabian Peninsula and North Africa (Banerjee et al. 2019). Due to its high prevalence in dromedaries, the virus will continue to spread to animal owners and caretakers as well as tourists. The case-fatality rate is even higher than SARS, currently at 34.3% (WHO 2020b). Although the disease tends to be sporadic, MERS should not be underestimated. Some researchers suggest that MERS is currently in what might be called the "calm before the storm" phase. Provided the virus acquires the ability to be effectively transmitted human-to-human through random genetic mutations, we may be in for the next coronavirus pandemic (Graham et al. 2013; Corman et al. 2018).

**COVID-19** (derived from the English name "coronavirus disease 2019") was first detected in the Chinese city of Wuhan in late December 2019. The pathogen, **SARS-CoV-2,** has some similarities to the SARS virus and causes lung disease, some of which is severe. Analysis of the genetic information of the new virus suggests that it has been circulating in humans since at least early November 2019 (Li et al. 2020). The virus spread around the world in a matter of weeks and months and was classified as a **pandemic** by the WHO in March. Currently, over 120 million infections and nearly 2.7 million deaths have been confirmed worldwide (CSSE 2021). In this disease, in addition to the severe courses and lethality, especially for the elderly or people with previous disease, there are also unusual late effects, with terms such as **long covid** or **post covid,** which are unusual in this frequency and intensity. This includes disease symptoms such as long-term reduced ability to perform or concentrate. Even severe neurological damage is possible, the exact extent and severity of which is being intensively researched (Wang et al. 2020).

The virus particle, called a **virion,** consists of only four proteins, a viral envelope, and the viral genome (Fig. 1.2). On the outside is the viral envelope, a

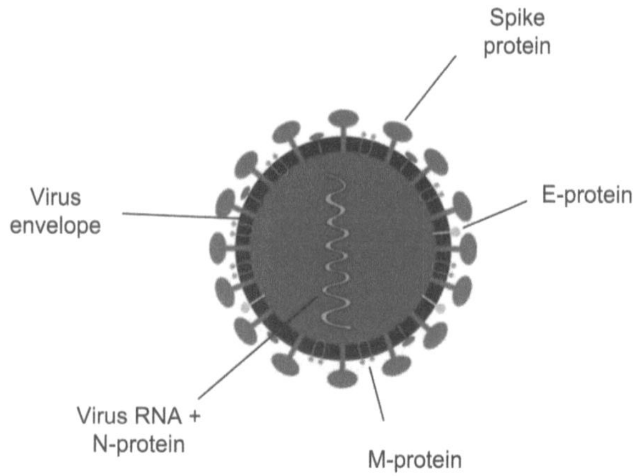


**Fig. 1.2** Structure of the virus particle. (Source: Adobe Stock, file no.: 339973957, modified)

membrane that the virus obtains from the last infected cell by budding from the surface. The so-called **spike protein** forms the long, name-giving projections with which the viruses dock onto cells. It is of particular importance for vaccine development because it activates the immune system particularly well. The other two surface proteins have a supporting function, for example in the assembly of new virus particles. Inside the virus is the genetic information in the form of a ribonucleic acid (RNA), also known as the viral genome. This is encased and stabilised by the fourth so-called N protein (Fehr and Perlman 2015).

The four proteins that are part of the **virion** are called structural proteins because they form the finished virus particle. In addition to these, the viral genome also contains the genetic information for other proteins. These are formed only after infection of a cell and help to multiply the virus. These proteins, or more precisely enzymes, produce large amounts of messenger RNA (for the subsequent formation of viral proteins) and further copies of the viral genome. This leads to the formation of many progeny, i.e. complete virus particles, in virus-infected cells. The new virions are assembled in the cell and released at the cell surface. In addition, there are so-called auxiliary proteins that help the virus to disrupt or deceive the host's immune system (de Wit et al. 2016).

# 2 Vaccine Technologies, Approaches Against COVID-19, Clinical Phases

## 2.1 Overview of Vaccine Technologies

Shortly after the enormity of this **pandemic** became clear and local containment was illusory, unprecedented preparations to develop vaccines against **COVID-19** began. The scientific community and the pharmaceutical industry are focusing on a broadsword approach, with particular emphasis on newer technologies. **WHO** is currently systematically recording all ongoing COVID-19 vaccine development projects and their progress (WHO 2020a).

Figure 2.1 summarizes the different approaches to vaccine development that are available today against infectious diseases and that we will learn about in this book. One approach is to modify the virus to make it less dangerous. This type is called a **live vaccine** and then contains whole virus particles capable of replication (Chap. 3). The second classical approach is to inactivate the virus, e.g. by chemical reagents. This **inactivated vaccine** also contains whole virus particles which, however, are no longer able to replicate. Further developments of this are **split** and **subunit vaccines**, in which only parts of the inactivated virus are contained in the vaccine (Sect. 4.1). In **vector vaccines**, only a part, e.g. the genetic sequence for a protein of the dangerous virus, is introduced into a harmless virus. The vector helps to deliver the target sequence into the cells of the body where the viral protein is produced (Sect. 3.3). In contrast, **virus-like particles** are empty shells or particles that have no viral nucleic acid inside (Sect. 3.4). Two technologies that are also new are not aimed at using another virus, but at delivering genetic information directly into cells in the form of nucleic acids. These include **DNA** and **mRNA vaccines** (Chap. 5). Another approach is based on **recombinant proteins,** also called

© The Author(s), under exclusive license to Springer Fachmedien Wiesbaden GmbH, part of Springer Nature 2023
P. U. B. Vogel, *COVID-19: Search for a vaccine*, essentials,
https://doi.org/10.1007/978-3-658-38931-4_2

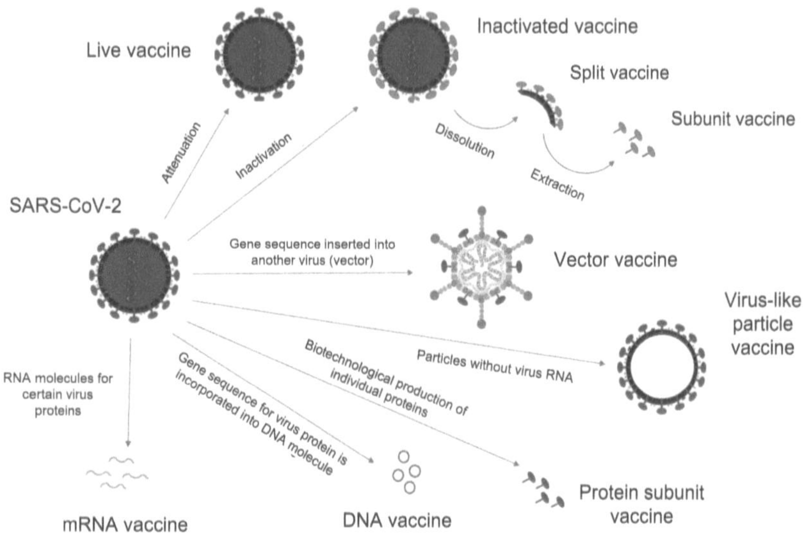


**Fig. 2.1** Overview of different vaccine types. (Source: Created using and modifying Adobe Stock file no.: 339973957 and 350847731)

subunits. The proteins are produced by genetic engineering methods in, for example, bacteria and then administered as protein vaccines (Sect. 4.2).

There are about 240 different vaccine candidates in the running, of which about 40% are already in the **clinical phases** (WHO 2020a). In terms of numbers, the projects are led by protein-based candidates (approx. 1/3 of the projects) and vector viruses (approx. 20% of the projects), followed by nucleic acid-based candidates, while the classical approaches (live vaccines and inactivated vaccines) together account for less than 10%. Furthermore, of these clinical-stage vaccines, several have already received regulatory approval (as conditional, emergency or regular) in various countries, including mRNA, vector and inactivated vaccines (vfa 2021), among others.

## 2.2 Immune Response and Important Concepts

What happens during a vaccination? The human immune system consists of two so-called arms, the **innate,** and the **adaptive immune response** (Müller et al. 2008). The innate immune response is non-specific and kicks in when a pathogen enters the body, giving us early general protection, so to speak, but in some cases is not sufficient to prevent disease. The adaptive immune response is based on specific recognition of a pathogen and leads, for example, to the formation of large amounts of antibodies (humoral immune response) that coat and thus block the virus, for example, and/or immune cells that attack virus-infected cells, for example (cell-mediated immune response). The starting point are so-called **antigen-presenting cells,** e.g. phagocytes, which patrol the intercellular spaces in our body as "guards". When they find viruses or their components, they "swallow" them, process them into smaller fragments and display these fragments on their cell surface. They migrate to small lymph nodes, places where large numbers of immature immune cells are virtually waiting for the order to deploy, and activate them. As a result, specialized immune cells multiply and differentiate and begin to fight the new virus. These processes also take place in the body after a vaccination, i.e. as a reaction to the components contained in the vaccine. The decisive advantage now is that some of these armed immune cells temporarily retire as so-called memory cells. If the body comes into contact with this virus again after the vaccination, it has direct protection through e.g. antibodies. Furthermore, the dormant immune cells immediately multiply and prevent disease from occurring.

If a vaccine candidate causes high antibody levels in humans, this is a measure of the **immunogenicity** of the vaccine, i.e. a measurable immune response is produced in the recipient. However, this is not necessarily the same as **efficacy**. Efficacy refers to how well one is protected against the particular disease. While for many infectious diseases a high antibody level is equated with protection against the disease (e.g. measles), there are other diseases for which antibodies are not suitable for inferring protection. A good example is the respiratory syncytial virus, which causes a febrile cold predominantly in young children, but also in adults. Here, despite having overcome the illness and having high levels of antibodies in the body, a person can fall ill again, although the course of the illness is milder in such cases (Piedimonte and Perez 2014).

In the case of animal vaccines, **efficacy** can be directly demonstrated experimentally. One group of animals is vaccinated, a second group remains unvaccinated, and then both groups are administered the disease-causing virus, called **challenge.** A direct comparison shows that the vaccinated animals are protected from disease, while the unvaccinated animals become ill. The Phase III studies, among others, to demonstrate the efficacy of the **COVID-19 vaccine candidates** were carried out without challenge. Groups were formed (vaccination and placebo) and monitored for signs of the disease over several months. Evidence that more cases of disease occurred in the unvaccinated placebo group than in the vaccinated group on a percentage basis demonstrated that the vaccinated group was better protected.

The second aspect is **safety.** This refers to all adverse reactions occurring in connection with vaccination, such as reddening of the skin, pain, fever, malaise or allergic reactions, but also any kind of secondary damage. Vaccines are the only medicines that are regularly used on healthy individuals across the board and should therefore cause as few adverse reactions as possible.

## 2.3 Preclinical and Clinical Phases

What is actually the difference between **preclinical** and **clinical?** The development of a vaccine begins with the identification of a suitable candidate, e.g. a virus isolate or antigen that is recognised by the immune system and induces a strong immune response, e.g. the spike protein of **SARS-CoV-2.** The preliminary construction of the vaccine candidate is followed by various preclinical tests, e.g. tolerability, immunogenicity and efficacy studies in animals. Significant requirements must be met before a candidate is "unleashed" on humans (Schriever et al. 2009; Pfleiderer and Wichmann 2015). The next stage is clinical phases I to III and, after successful approval, further observation of the safety and efficacy of the new product in the market (clinical phase IV):

- **Preclinical phase:** identification of the antigen, construction of the vaccine candidate, tolerability and efficacy studies in animals
- **Clinical Phases:** Use of the vaccine in humans to test safety, tolerability, dose finding, and efficacy with progressively increased numbers of subjects from Phase I to III (pre-approval) and Phase IV (post-approval).

**Phase I** essentially serves the **proof-of-concept,** i.e. the proof that the candidate basically works as expected. The aim is to test safety and tolerability, but also immunogenicity, on a small number of test subjects (e.g. 30–50). The test subjects are usually given intensive medical care at the beginning in order to be able to react quickly to unexpected complications. The importance of this phase is illustrated by a serious incident. In 2006, a biological product, an antibody called **TGN1412,** was tested on humans for the first time. Here, after administration of small doses to six test subjects, completely unexpectedly, severe side reactions up to coma occurred. This incident showed that one cannot necessarily conclude from the tolerance in animals to the tolerance in humans. The next stage is then **clinical phase II,** which is used for **dose finding.** It is proven that the intended dose (amount of viral particles or molecules administered to the recipient) causes a sufficient immune response and is safe. This phase is done on a larger number of subjects, typically 200–400. The **phase III clinical trial** is then the last large-scale trial before approval. A "promising" candidate has been found and its suitability is now tested on thousands of subjects. In the case of the clinical trials of the **COVID-19 vaccine candidates,** a large number of subjects were included, which improved the interpretation of the results. Once the vaccine has been approved, further evaluation regarding safety and efficacy will take place, called **clinical phase IV** (Volkers et al. 2005).

Why, if the previous phases were successful? The clinical phases I–III are only samples, i.e. tests on a few thousand people in total. If a new vaccine causes a serious secondary disease at a very low rate, this will be difficult to detect in **clinical phases I–III**, but more likely to be detected if the vaccine is used routinely in hundreds of thousands or millions of people. This is why the use of new licensed vaccines is monitored particularly critically by the competent authorities in the first few years. This is particularly important for **COVID-19 vaccines** because of the recent launch of an unprecedented vaccination campaign in which every batch of vaccine produced is being vaccinated in record time.

## 2.4 Current Status

Fortunately, none of the previously possible risks have materialized or torpedoed the progress of vaccine approvals. In the early stages of vaccine development, there were concerns that **COVID-19 vaccines** could cause an amplification of disease progression following subsequent infection, as has occurred e.g. with **SARS vaccine candidates** in some animal models (Tseng et al. 2012). Based on these concerns, recommendations have been developed by various organizations that should be considered in this regard during vaccine design and subsequent phases (Lambert et al. 2020). The basic concerns could be mostly addressed as early as 2020. For example, in animal studies, an **adenovector virus** (type Ad26) was effective and safe without causing any damage in rhesus monkeys after the so-called **challenge,** i.e. infection with a virulent strain (Mercado et al. 2020). Similarly, in hamsters, when SARS-CoV-2 antibody-containing plasma was transferred to animals and subsequently infected, no adverse effect was also observed in some COVID-19 patients (Wen et al. 2020). Although not all studies and candidates can be listed here, in principle, an analysis of clinical and biological characteristics also concluded that this was unlikely in COVID-19 (Halstead and Katzelnick 2020). For example, another risk was skipping **phase III clinical trials** in some regions of the world due to enormous public, economic and political pressure. In this regard, the **Paul Ehrlich Institute** had already clarified in the first half of the year that the usual requirements must be fully met. The PEI is also involved in centralized approvals, which in Europe are handled by the **EMA,** and is given full access to the documents. Ultimately, the pharmaceutical companies and regulatory authorities took the time necessary to bring the first vaccine candidates to **regulatory readiness** or to be able to assess the **quality**. Considering that various agencies such as the FDA and WHO set vaccine efficacy expectations at 50% and 60%, respectively, early in 2020 to pave the way for first-generation vaccines, this "understatement" was not at all necessary. The efficacy data were impressive, led by the first communicated data from the **BioNTech/Pfizer** and **Moderna** mRNA vaccines, but also, for example, the **Oxford University/AstraZeneca** vector vaccine, which was the first vector vaccine to receive approval in the EU, with a combined efficacy of just over 70% (Voysey et al. 2021a). The safety data here were also good. For example, although a wide variety of **adverse events** occurred with the AstraZeneca vaccine, most were not attributable to the vaccine (also more adverse events in control group than in vaccinated group) and only a few severe adverse events were presumably attributed to the treatment (one of the three cases also occurred in the control group). The affected subjects recovered completely or are still recovering. However, a limitation was the small number of subjects over 70 years of age (Knoll and Wonodi 2021). For this reason, the Standing Commission on Vaccination in Germany, for example, recommended limiting use to people up to 64 years of age. However, this restriction was lifted again in March 2020 on the basis of further data, so that all people aged 18 and over can be vaccinated with this vaccine.

**Side effects** are an important issue with vaccines. No vaccination is absolutely free of side effects, but serious side effects are very rare (Dittmann 2002). A vanishingly small proportion occur immediately, e.g. in the form of allergic or severe **allergic reactions.** The majority of side effects occur in the hours and days following vaccination, such as pain at the injection site or fever. Again, a vanishingly small proportion may show up weeks or months later, for example. With conventional vaccinations, which have been used for a long time, the rate of severe immediate reactions is very low. For example, with the **MMR vaccine** (measles, mumps, and rubella), there is a rate of severe allergic reactions of less than 1: 1,000,000; with other adverse reactions, the rate may be higher in some cases (Spencer et al. 2017). An early evaluation of allergic reactions including anaphylactic shock following **COVID-19 vaccination** in America found a rate of approximately 11 cases per million vaccine doses after nearly 2 million vaccinations, with a median reaction after 12 min. In many cases, the affected individuals had a history of allergy (CDC 2021). Therefore, the recommendation in Germany to stay at the vaccination center for 30 min following vaccination is also understandable, so that one has direct medical attention in case of allergic reactions. In addition, the **PEI** has provided an app for reporting side effects after COVID-19 vaccinations, in which affected persons can report side effects directly. After the start of vaccinations also in GP practices, this is an important innovation to shorten the usually well-functioning reporting chain (doctor → public health department → PEI) and to be able to react quickly should any irregularities become apparent.

The authorities intensively monitor the **side effects** of vaccinations. In Germany, the **PEI** publishes regular updates with statistics. Currently (as of 12.02.2021), just under 4 million vaccinations have been carried out with the three vaccines already approved (BioNTech, Moderna, AstraZeneca). The average rate of general adverse events is 1.9 per 1000 vaccine doses and of serious adverse events 0.3 per 1000 vaccine doses. Overall, the PEI thus confirms no unusual incidence and a continued positive benefit-risk profile. Specifically, side effects of the AstraZeneca vaccine are predominantly limited to local (arm pain) or flu-like symptoms after administration of approximately 3 million vaccine doses in the UK, which is very positive (PEI 2021). However, this vaccine subsequently showed a slight cluster (less than 1 case per 100,000 vaccinated) of certain thromboses, some of which were fatal. This led to a temporary suspension of vaccination in several European countries. However, causality has not yet been confirmed and both WHO and EMA confirmed a continued positive risk-benefit ratio following the March 2021 investigation.

A final aspect of **side effects** is the question of other **consequential** or **long-term damage.** Consequential damage may take the form of chronic or transient disease, and often causality (independent of vaccination or resulting from vaccination) cannot be proven. For example, the MMR vaccine is still associated with **autism,** although various evaluations show no evidence of this (Aps et al. 2018). This assumption was based on a scientific study, but it was falsified (Spencer et al. 2017). A more common example is **Guillain-Barré syndrome** (GBS), a disease with paralysis symptoms from which sufferers often recover. This condition is known to occur with influenza vaccinations, although the rate is much more common after natural influenza infections. Overall, simulations suggest that vaccination campaigns are more likely to reduce the incidence of GBS in the population (D'alò et al. 2017). However, a possible association also seems to depend on the vaccine, e.g. no clustered incidence with seasonal vaccines, but somewhat increased with earlier pandemic vaccines (Prestel et al. 2014). Regarding **COVID-19** and GBS, there are at most isolated cases, but not yet enough data to draw reliable conclusions about whether the disease is truly associated with this syndrome (Zhao et al. 2020). However, GBS often manifests in the first weeks after infection or vaccination, and against the background of more than 100 million COVID-19 cases worldwide and many millions of vaccinations, a real association should already have become more apparent, especially because medically very close attention is paid to **long-covid symptoms.** The current data on possible late effects of COVID-19 vaccination is reassuring, as most late effects are expected in the first 2 months after vaccination (Kim et al. 2021). Since many tens of thousands of people have already participated in clinical trials in the past year and millions of vaccinations have already occurred in December 2020, this critical period has passed. Therefore, clustered sequelae are already unlikely. Nevertheless, there remains a need for transparent data collection, communication and accurate analysis (Hampton et al. 2021).

# 3 Live Vaccines, Vector Vaccines and Virus-Like Particles

## 3.1 Classical Live Vaccines

**Classically attenuated live vaccines** were among the first vaccines developed against infectious diseases. This type of vaccine dates back to early research at the end of the eighteenth century. Although the principle had been practiced earlier, an English physician, **Edward Jenner**, was the first to publish the findings in a scientific journal. He administered a cowpox preparation to a boy and later infected the boy with the smallpox virus, which was dangerous to humans. The boy was protected from contracting the disease by this pretreatment. This scientific work is considered the birth of **modern vaccination** (Riedel 2005).

The idea of live vaccines is to attenuate a virus in its **virulence** so that it no longer causes disease, but is recognized by the immune system to trigger a protective immune response. The weakening of virulence is called **attenuation.** The pathogen is thus quickly controlled and eliminated by the immune system. Thus, the damage from the vaccine virus is either non-existent or greatly attenuated. Live vaccines have historically been among the most effective types of vaccine, as they mimic the processes of a natural infection and have a strong stimulating effect on the immune system by multiplying in the target tissue. They also usually confer particularly long-lasting immunity. **Live vaccines** contributed to the eradication of smallpox and are still standard against many infectious diseases, such as measles, mumps and rubella (Minor 2015).

An important feature here is their **empirical development** according to the **trial-and-error principle,** whereby the molecular basis of this attenuation remained obscure. Two classical coronavirus vaccines against infectious bronchitis of poultry (see Chap. 3) are good examples to understand the empirical process of

© The Author(s), under exclusive license to Springer Fachmedien Wiesbaden GmbH, part of Springer Nature 2023

P. U. B. Vogel, *COVID-19: Search for a vaccine*, essentials,
https://doi.org/10.1007/978-3-658-38931-4_3

attenuation and the relationship between the important concepts, efficacy and safety. These strains, called H52 and H120, were developed from the 1950s and subsequently licensed as vaccines by various companies, so have been in use for over half a century. Strain H120 is still among the most commonly used vaccines today (Ramakrishnan and Kappala 2019).

The development of these vaccines goes back to the Dutch virologist **Gosse Bijlenga.** In 1954, an outbreak of infectious bronchitis occurred on a poultry farm in the Netherlands. To create a vaccine, Bijlenga used samples from sick animals on this farm. The scientist injected the aggressive virus isolate into the albumen of chicken eggs and incubated them. After a few days, he removed the egg white from these eggs and injected it again into the egg white of new chicken eggs (Bijlenga et al. 2004). The idea behind this was to continuously adapt this aggressive virus to the embryonic tissue of the chicken embryo by multiplying it in the chicken egg and thereby weaken its virulence for adult tissue (chicken).

In total, Bijlenga made this propagation step in chicken eggs 52 times (= H52, the H stands for the surname of the owner of the farm from which the isolate came). He then tested virus material from the 25 passage (H25) in direct comparison to H52 under controlled conditions on poultry farms. The H25 strain was far too aggressive and unsuitable as a vaccine. The animals vaccinated with the H52 strain showed a strong **antibody response,** but the strain was still too aggressive, especially in younger chickens. For this reason, further passages were made in chicken eggs up to passage 120 (Fig. 3.1). This preparation was clearly safer to use also on younger animals, but still stimulated a strong **immune response** (Bijlenga et al. 2004).

The properties, **efficacy** and **safety,** can be roughly recorded in a diagram (Fig. 3.2). From this follows the general principle: the further a virus is removed from its actual host, the safer the application becomes, but the more likely it is that the virus will be "overweakened", i.e. will no longer exhibit the hoped-for efficacy.

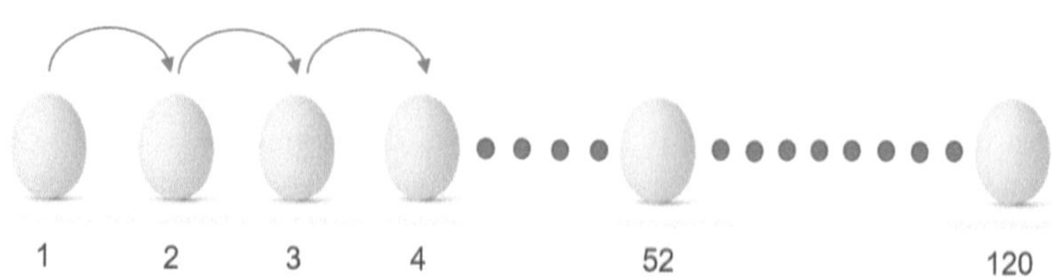


**Fig. 3.1** Schematic representation of the development of the live attenuated vaccines H52 and H120 against infectious bronchitis of poultry. (Source: Created using and modified from Adobe Stock, File No.: 76266497)

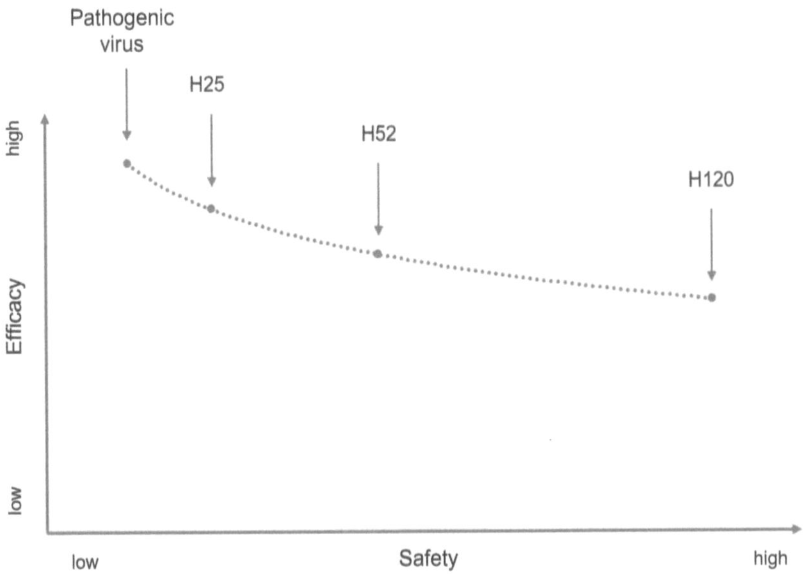


**Fig. 3.2** Schematic relationship between efficacy and safety in the stepwise attenuation of a virus

## 3.2 New Approaches Live Vaccines

Although these vaccines are still in use in many areas and serve us well, attempts are now being made to develop live vaccine candidates no longer according to this classical principle. Technological advances give researchers the ability to make targeted changes to viruses. One approach is called **codon deoptimization.** What does this mean? The **genetic code** consists of a sequence of nucleic acid triplets. Each gene consists of a chain of nucleic acid triplets, each coding for an amino acid. This information is passed on to mRNA and converted into proteins at ribosomes. Here, however, the genetic code is degenerate, meaning that certain amino acids are encoded by multiple nucleic acid triplets (called **codons**). However, this usage is not random or universal, and different organisms prefer certain codons, including our body cells. Example: 4 different codons, including CUG and CUA, encode the same amino acid valine. Our body cells "love" CUG, meaning that in our gene sequences for proteins containing valine, CUG is very often found. The codon CUA is used rather rarely. The strategy now is to replace frequently used codons with rarely used codons by selectively changing the viral genome, but without changing the amino acid. Provided that this is done at different sites, the genetic information of the viruses can still be read in our cells and converted into protein, but at a considerably lower rate. Thus, it is not the virus itself that is altered, but the rate of protein synthesis (Le Nouën et al. 2019). However, the right mix of **attenuation** and **immunogenicity** must be found virus-specifically (Meng et al. 2014).

According to current **WHO** information, three projects (rounded 1%) on live attenuated vaccines against **COVID-19** are being pursued, all based on targeted codon usage modification (WHO 2020a). One candidate, **COVI-VAC** from Codagenix in collaboration with the Serum Institute of India, is in early clinical phase (Cision PR Newswire 2020). The reason for this secondary consideration is that one needs to know virus replication and genetic stability in great detail. Especially with highly pathogenic viruses, live vaccines are particularly risky because you are "releasing" the vaccine specifically and on a large percentage of the population. A vaccine that has been attenuated in its virulence by, for example, passages in cell culture or by genetic engineering methods, has the potential to regain all or part of its virulence during replication in the vaccinated individual. This phenomenon is referred to as **"reversion to virulence".** An example of this is cases of poliomyelitis after vaccination with the live polio vaccine (Minor 2015). For this reason, i.e. a very rare occurrence of vaccine-induced severe disease, **oral polio vaccination** (live vaccine) has been replaced by an inactivated vaccine in Germany, for example (Zündorf and Dingermann 2017). In addition, there is the even more common problem that live vaccines are not always well tolerated by immunocompromised individuals and the very elderly, among others.

In addition, it is known from some patients with severe **COVID-19** that autoreactive antibodies can form, i.e. antibodies directed against the body's own structures (Pharmazeutische Zeitung online 2020), which must be ruled out for the vaccine. Therefore, it is particularly important here that the different clinical phases are not carried out in parallel, but one after the other. This is also taken into account by the manufacturer of **COVI-VAC**, who is only planning further clinical phases later after the first results of clinical phase I are available in mid-2021 (Cision PR Newswire 2020).

If one of the **live vaccines** should prevail, one advantage could be the effectiveness against **virus variants.** Live vaccines have many different antigenic structures, so-called **epitopes,** against which the immune system builds up a protective immunity. Therefore, the risk that the efficacy is undermined by viral mutations is lower on average.

## 3.3 Vector Vaccines

The risks of live vaccines have long driven the search for alternatives. One approach, **vector vaccines,** aims to eliminate or minimize these risks. In this technology, one does not want to use the whole virus, but only a part, e.g. the important spike protein of **SARS-CoV-2.** For this, one needs a transport vehicle, the so-called **"vector".** These vectors are themselves viruses whose molecular biology is well studied and which do not cause disease. The additional protein may be directly incorporated into the surface of the viral particle or may only lie dormant as a genetic sequence in the viral genome (Vujadinovic and Vellinga 2018). In the second case, the protein is only produced by our body cells once the vector virus has entered the cells. This type of vaccine is specifically designed on a computer and then "assembled" in the laboratory using **genetic engineering methods.** Of course, these also need to be empirically tested in animals and humans to see if they are suitable as vaccines, but the design of the vaccine is a very targeted, deliberate process.

According to WHO, various vector viruses are currently being tested as vaccine candidates (WHO 2020a). Here we will get to know some selected vectors that have entered the race against **COVID-19.** Two different types of multiplication ability are distinguished (Rauch et al. 2018):

- **Propagating vectors:** After administration, these can multiply in the first infected cells and subsequently form infectious virus particles again, which infect further body cells. During each cycle, for example, the spike protein is also formed (either as a single protein or as part of the surface of the virus particle), which activates the immune system. However, the multiplication rate of these vector viruses is so slow that they do not cause any harm
- **Vectors incapable of reproduction:** The virus particles contained in the vaccine can infect cells. Certain viral proteins are formed in these cells, but the viruses are no longer able to form new infectious virus particles due to genetic defects.

An important advantage of **vector vaccines** is that they partially mimic a natural infection and activate both the formation of antibodies and the cell-mediated immune defense. Another positive aspect of this technology is that it is already partially proven in the market. There are about a dozen approved vector vaccines in the veterinary field and also the first vaccines in the human field. These include, for example, the first vaccine against Ebola, rVSV-ZEBOV, which received approval from the **European Commission** in 2019 (PEI 2019). In addition, dozens of projects testing RNA vector viruses against a wide range of infectious diseases are currently in the preclinical phase (Lundstrom 2019).

There are various viruses that can be used as vectors, such as **adenoviruses,** the vesicular stomatitis virus (VSV), or the "Modified Vaccinia Virus Ankara" (MVA). Some approaches also combine vector viruses with antigen-presenting cells that present **SARS-CoV-2 antigens** to stimulate the immune system. The different approaches each have their own advantages and disadvantages. Adenoviruses are the most commonly used vector viruses to date; according to WHO, most vector virus projects against **COVID-19** are based on adenovirus vectors (WHO 2020a). This is partly because the virus properties are the best researched and **adenovirus vector technology** is the most advanced overall. For this reason, it is not surprising that these types were among the first vector vaccines to be licensed. One disadvantage of **vector vaccines** could be **prior immunity** to individual vectors. For example, 40–45% of people in the United States have antibodies to adenovirus 5, the most common type in humans (Saxena et al. 2013). Some of the vector viruses used could thus be blocked by the immune system before they take effect (Awadasseid et al. 2021). However, this needs to be clarified in more detail, as these vaccines contain an incredibly high amount of virus particles that could also "override" pre-immunity.

In August 2020, the world's first **COVID-19 vaccine** received emergency approval in Russia. This vector vaccine, called **Gam-COVID-Vac** (or Sputnik V), comes from the Gamaleya Research Institute. This was followed by considerable criticism, as hardly any results were available, i.e. the efficacy and safety could hardly be publicly assessed. In September, results of the clinical phase I/II were first published publicly, but this did not allow the criticism to subside (Balakrishnan 2020). Meanwhile, very good preliminary **efficacy data** of about 91.6% were published in January 2021 (Logunov et al. 2021). The vaccine is already in use in several countries and approval in the EU is also being sought. In light of the supply shortages experienced by manufacturers in the spring of 2021 and the continued possibility of supply shortages occurring again, this, along with other candidates, could ensure **vaccine supply**. This vaccine uses a combination of two different adenoviruses (rAd5 and rAd26) to prevent vaccine blockage by possible prior antibodies following natural adenovirus infection. This is important as there are large regional differences in the seroprevalence of certain adenoviruses, e.g. North America and Europe have high Ad5 seroprevalence (Mennechet et al. 2019).

The **vector vaccine AZD1222** from Oxford University in collaboration with AstraZeneca was the first to receive conditional approval in the EU. The major advantage of this vaccine is that large quantities can be produced due to its production capacity. In addition, the vaccine is less expensive than **mRNA vaccines** and can be stored at 2–8 °C. For this reason, this vaccine was considered to be particularly important for the EU in terms of market supply. Together with the other candidates, this **vaccine** is one of the great hopes for controlling the **COVID-19 pandemic**. In the meantime, the lower efficacy compared to other vaccines, which adds up to about 70%, was additionally presented negatively. It was noted here that this could lead to a two-tier society. In the end, we must remember the demand of science and politics. Achieving the important milestone of **herd immunity** is seen to be around 70%, so that the spread of **SARS-CoV-2** can be sufficiently stopped. Provided we hypothetically screen out all other vaccines and the entire population (with children not currently being vaccinated) is vaccinated with AZD1222, at least theoretically (since not all adults will get vaccinated) this necessary 70% immunity rate would be achieved in the adult group. In addition, this efficacy value is really good, as it is on average higher than for influenza vaccines (Bouvier 2018). Further, AstraZeneca has a big advantage. The sometimes contradictory clinical trials suggest that without changing the product itself (which is very time consuming), it is possible to improve **efficacy** by changing the application, for example, like the dose used in the first application. In addition, extending the interval between the first and second application also shows positive effects (Voysey et al. 2021b).

The vector vaccine **Ad26.COV2.S** from Janssen (Johnson & Johnson) has also received conditional approval in the EU. According to interim results, the efficacy is 66%, which is similar to that of the first vector vaccine approved in the EU. A major advantage is that only one administration is required. This will facilitate and accelerate vaccination campaigns if sufficiently available.

In principle, the **safety profile** of **vector viruses** is more favourable than that of classical live vaccines. However, one safety aspect that is repeatedly discussed, especially in the case of DNA viruses, is the risk of accidental insertion of genome parts of vector viruses into the **human genome.** Here it is important to remember that this is a theoretical risk. When we look at the human genome, many fragments of viral nucleic acids are found. In total, it is estimated that 8% of the human genome is of viral origin. These "remnants" of various viruses are witnesses to millions of years of evolution and have not permanently harmed us. In addition, **adenoviruses** are common causative agents of rather mild colds in humans, among other things. Almost all children up to the age of five, as well as adults, go through adenovirus infections (Heim 2016). During the cold season, adenovirus infections account for up to 5% of common colds, depending on the region (Heikkinen and Järvinen 2003). In addition, there are also adenoviruses that persist for years in body cells (e.g., the rachial tonsils or lymphoid tissues), meaning that their DNA remains in the nucleus for years. These persistent infections have never been confirmed to be associated with causing tumors (Heim 2016). There are some viruses that can trigger tumors, such as human papillomaviruses, but adenoviruses are not counted among the viruses with these tumor-initiating properties (Morales-Sánchez and Fuentes-Pananá 2014; Gaglia and Munger 2018). Actually, accidental insertion of the genome should not be a serious problem both in vaccinations with adenovirus vectors and in natural adenovirus infections, since the infected cells produce viral protein and thus become themselves the target of the immune system attack, ultimately leading to their destruction. So one should always balance this theoretical risk against the real risk of **COVID-19 disease,** including long-term damage such as certain neurological damage in **Long-Covid,** which is many orders of magnitude more likely. On the other hand, very rarely occurring side effects such as spinal meningitis are known (Kremer 2020), from which those affected usually recover. The currently occurring slight accumulation (less than one case per 100,000 vaccinated persons in Germany) of certain thromboses with the AstraZeneca vaccine, which have led to a temporary vaccination stop in various countries, are also being investigated very closely. However, it must also be borne in mind that no other vaccine is absolutely free of side effects (Dittmann 2002).

A problem with regard to a high **willingness to vaccinate** is when theoretical risks (e.g. DNA insertion) manifest themselves in people's minds as a high health risk. Then it becomes difficult to regain confidence. This theoretical risk also exists for pure DNA vaccines, but is considered to be very low (see Sect. 5.1). Finally, perhaps a comparison with another area will help. Airbags are used in road traffic as a life-saving measure (analogy: vaccination) in the event of an accident (analogy: COVID-19 infection). Nevertheless, there is a minimal theoretical risk of being injured or dying due to a misfiring airbag. Would we therefore, out of fear of this theoretical risk, have the airbag removed or only buy cars without airbags (analogy of not getting vaccinated)? Probably not, and that's why no one should be afraid of **vector vaccines**. It can be assumed that this type of vaccine, together with other **vaccine technologies,** will become more and more widespread in the coming years and decades and will become a "standard vaccine".

## 3.4 Virus-like Particles

Unlike vector vaccines, which use a virus as a transport vehicle for sequences or proteins of **SARS-CoV-2**, virus-like particles (**VLPs**) are based on the administration of an empty virus-like envelope, i.e. the particles do not contain nucleic acid (Crisci et al. 2013). These vaccines are therefore not capable of replication and cannot form new viral proteins in somatic cells. The proteins from SARS-CoV-2 are directly incorporated or bound into the particles to elicit an immune response. There are several approaches, ranging from constructs that contain a lipid shell with embedded proteins to constructs in which individual proteins assemble into a stable particle (Chroboczek et al. 2014; Syomin and Ilyin 2019). One of the systems commonly used for VLPs is the so-called baculovirus expression system. Here, genetic engineering methods are used to introduce the desired sequences for specific proteins of **SARS-CoV-2** into the so-called baculovirus, which is subsequently propagated in cell cultures, e.g. in bioreactors. However, only the released proteins that assemble into particles are harvested and processed into the vaccine (Felberbaum 2015).

This technology has celebrated some successes in recent years. Similar to vector vaccines, there are approved **VLP-based** vaccines for both animals and humans (Crisci et al. 2013; Felberbaum 2015). An example is several vaccines against human papillomaviruses, which cause mostly benign tumors on skin and mucosa, or a hepatitis B vaccine (Mohsen et al. 2017). The great advantage of this approach is its **safety,** no replicable viruses are formed, with simultaneous high **immunogenicity.** Due to its characteristics, the technology has received a positive evaluation by the Paul Ehrlich Institute, which itself is researching further improvements to this technology (PEI 2018). There is also the possibility of using VLPs as a multivalent vaccine. This type is based on particles containing proteins from different pathogens. For example, the suitability of a VLP vaccine against several mosquito-borne diseases, including Zika and yellow fever, is currently being tested (Garg et al. 2020).

According to **WHO data,** almost 20 **VLP-based** vaccine projects are currently underway against **COVID-19** and the first have advanced to the clinical phases (WHO 2020a). Even though this technology has already proven that it is not only theoretically but also practically suitable as a vaccine, this cannot be generalised for every infectious disease. The most advanced candidate, **Coronavirus-like Particle COVID-19,** developed by Medicago in cooperation with GSK, is already in combined phase 2/3 after good results in the first clinical phase. A special feature of this approach is that the vaccine is produced in plant cells. It is to be hoped that these projects will be successfully completed and quickly complement the existing vaccines.

# 4 Inactivated Vaccines and Recombinant Protein

## 4.1 Inactivated Vaccines

**Inactivated vaccines** historically form the second major vaccine group, the antithesis of live attenuated vaccines, so to speak. In inactivated vaccines, the virus is first multiplied in large quantities and then inactivated, e.g. by chemical reagents. As a result, the viruses lose their infectivity and can no longer reproduce. Three types are distinguished (Fig. 4.1).

Depending on the disease, all three types are still in use today. Against influenza, for example, there are various vaccine types worldwide, in addition to live vaccines also all three types of inactivated vaccines. Historically, the **inactivated whole virus vaccine** is the oldest type because it can be produced with relatively limited technical capabilities. However, high safety standards, biological protection level 3 in the case of **SARS-CoV-2**, must be maintained during manufacture. The production of the **split vaccines** and **subunit vaccines** is more costly due to the additional steps involved. The first subunit-based vaccine was licensed for hepatitis B in 1981. These three types each have certain advantages but also disadvantages. The advantage of whole-virus preparations is that they contain different viral proteins and components. Thus, when the vaccine is administered, a wide variety of **epitopes** (e.g., immunologically relevant regions of proteins) are presented to the immune system, against which the immune system establishes a specific response. Whole-virus preparations generally activate the immune system more strongly than split- or subunit vaccines, but also show on average more side reactions, such as fever or swelling. In Germany, for example, more than 10 different influenza vaccines are licensed, of which all inactivated types are based on split- or subunits (PEI 2020). Although there are differences within these types, the resulting

© The Author(s), under exclusive license to Springer Fachmedien Wiesbaden GmbH, part of Springer Nature 2023
P. U. B. Vogel, *COVID-19: Search for a vaccine*, essentials,
https://doi.org/10.1007/978-3-658-38931-4_4

- **Whole virus vaccine**: here the whole viruses are inactivated. The vaccine contains complete virus particles that are no longer infectious.

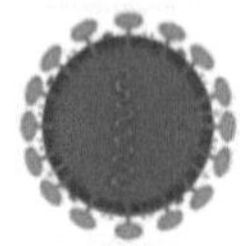

- **Split vaccines**: here the viruses are additionally dissolved after inactivation, i.e. the membrane is dissolved and the interior of the virus particle, i.e. the virus genome with the bound proteins, is also removed during purification. the vaccine then consists of membrane fragments with embedded viral proteins.

- **Subunits**: here the split vaccines are further processed, i.e. certain important viral proteins are extracted from the membrane and the membranes are removed, thus obtaining soluble proteins (subunit vaccines can also be produced by recombinant methods, see Sect. 4.2).

**Fig. 4.1** Overview of different types of inactivated vaccines. (Source: Created using and modifying Adobe Stock, File No.: 339973957)

immune response is usually weak. For this reason, boosting substances, known as **adjuvants,** are added to the vaccine to have a stimulating effect on the immune system. These adjuvants are usually oil-based or aluminum salts. These vaccines are almost always administered **parenterally.** Inactivated vaccines are now standard against many infectious diseases, such as tetanus, diphtheria, polio, hepatitis A or rabies (Zündorf and Dingermann 2017).

This "old" but proven technology has kept pace with newer technologies to some extent. Early on, several candidates were in the clinical phase. One of the first candidates **CoronaVac** from the company Sinovac, a whole virus preparation that is propagated in cell cultures and then inactivated and purified, was shown very early to be safe, immunogenic and effective in preclinical trials including mice and rhesus monkeys (Gao et al. 2020). The vaccine is in phase III clinical trials with efficacy data to date of 50–91% depending on the trial (vfa 2021) and has already received emergency approval in several countries. Another advanced candidate is, for example, **BBV152** from Bharat Biotech (Rostad and Anderson 2021). Several other candidates are also in advanced clinical phase or have received emergency approvals. An inactivated candidate for Europe could be the product **VLA2001** from the company Valneva, although approval, if the next phases are successful, is not in sight until late in 2021 (Balfour 2021).

## 4.2 Recombinant Proteins (Protein Subunits)

The **subunit vaccines** (hereinafter referred to as proteins) presented in Sect. 4.1 can be produced not only from inactivated viruses, but also with the aid of biotechnological methods (= recombinant). In this process, the genetic sequence for the desired protein or part of the protein is inserted into certain DNA molecules, so-called **plasmids.** Plasmids are independent genetic elements (ring-shaped DNA molecules) that can replicate in cells. Plasmids are also used, for example, by bacteria to exchange antibiotic resistance, which can lead to multidrug-resistant bacteria. These plasmid DNA molecules are now multiplied. This is done by introducing them into cells such as the classic baker's yeast, ***Saccharomyces cerevisae,*** or bacteria such as ***Escherichia coli***. These organisms divide quickly and can be multiplied in large masses, in so-called **fermenters.** These fermenters are large tanks filled with nutrient fluid in which the growth conditions can be optimally controlled. The plasmids, just like the DNA of the cells, are passed on to the daughter cells with each cell division. In the end, you have a large biomass of cells that all have plasmids with the information for, for example, the spike protein of **SARS-CoV-2.** Using the genetic information of the plasmids, this recombinant SARS-CoV-2 protein is formed in the cells. After fermentation is complete, the proteins are separated and purified from the cells and other components, and finally adjuvants are added. This means that although the production process of inactivated (Sect. 4.1) and recombinant proteins is completely different, both require adjuvants to activate the immune system sufficiently. Hepatitis B is an example of a recombinant **protein vaccine** that is regularly used in humans. One advantage of recombinant proteins is the lower **biosafety level,** as the virus does not need to be replicated.

According to the **WHO,** most vaccine projects against **COVID-19,** about one third, are based on protein vaccines (WHO 2020a). This underlines how quickly these established biotechnological processes can be applied to new viruses. Already 20 candidates are in various clinical phases. As a caveat, a single protein is not necessarily as effective against a disease as classical vaccines. Of course, there are particularly immunogenic regions of a protein, called epitopes, in every virus, but the holistic immune response against whole viruses is usually directed against different proteins and here again different epitopes. Recombinant proteins therefore tend to have lower immunogenicity compared to traditional whole virus vaccines

(Karch and Burkhard 2016). For this reason, improvements in this area are the focus of intensive research efforts. One way to improve the immunogenicity of protein vaccines is through **nanoparticle technology,** such as coating nanoparticles with the proteins (Pati et al. 2018). The news on the first candidate approved in Russia, **EpiVacCorona,** which is considered to have 100% efficacy, is very good, although the publication of the data is still pending. Russia has already announced mass production of this vaccine (ärzteblatt.de 2021). For the EU, another option is the product NVX-CoV2373 from the company Novavax, which is in the approval phase in the EU. This vaccine is based on a recombinantly produced protein with additives. Following very good preclinical results (Tian et al. 2021), the efficacy data from clinical trials are also very good at just under 90% (vfa 2021).

# Nucleic Acid-based Vaccines

# 5

## 5.1 DNA Vaccines

Research into **DNA vaccines** began several decades ago. The idea behind this is that one does not introduce replicable viruses but certain gene sequences from them into the body's cells. The production of these vaccines has parallels with recombinant proteins. The selected sequence for a protein of **SARS-CoV-2** is incorporated into **plasmids.** These plasmids are now introduced into bacteria, for example, which are then propagated by **fermentation** in large tanks. In the bacteria themselves, the plasmids multiply in turn, so that each bacterial cell contains several identical plasmids. At the end of fermentation, the bacterial cells are "cracked open" and the plasmids are "harvested," and all other components (e.g., the outer coat of the bacteria, proteins, and bacterial RNA and DNA) are removed by purification steps (Fig. 5.1). The remaining plasmids are then mixed with adjuvant, for example, and filled into vials. Despite some parallels with recombinant proteins, a different biomolecule is harvested here, not protein but DNA. With the exception of the adjuvants, the vaccine contains almost "naked" DNA, which is injected into humans or administered via needle-free systems (Rauch et al. 2018).

Mode of action: The body cells take up the plasmids. Based on the genetic information of the **SARS-CoV-2** protein stored in the plasmid, normal gene expression takes place in our body cells, i.e. from DNA via RNA to protein. Whereas with recombinant proteins, protein antigens are injected directly into the body, **DNA vaccines** "force" the body's cells to produce this viral protein themselves. Since only a part of the viral genome (e.g. one or a few proteins or protein parts) is contained in the plasmid, no virus capable of replication can be formed. Even though recombinant proteins and **plasmid DNA** are two different biomolecules, their

© The Author(s), under exclusive license to Springer Fachmedien Wiesbaden GmbH, part of Springer Nature 2023
P. U. B. Vogel, *COVID-19: Search for a vaccine*, essentials,
https://doi.org/10.1007/978-3-658-38931-4_5

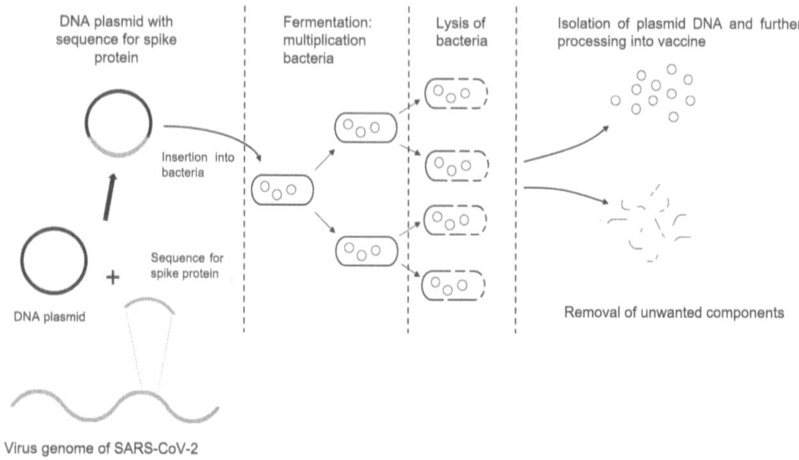


**Fig. 5.1** Schematic representation of the construction, propagation and subsequent purification of a DNA vaccine

pathway meets again at the stage when, for example, phagocytes present these protein parts (either injected or self-formed) to other immune cells and trigger an **immune response.**

DNA vaccine technology has evolved considerably since the initial phase, also with regard to the question of how the lack of **immunogenicity** can be improved by adjuvants. **DNA vaccines** are considered very safe to use as they cannot form replicable viruses. In a large number of >100 clinical trials, no meaningful harmful adverse reactions have been observed (Li and Petrovsky 2016). There are a handful of approved DNA vaccines in animals. So far so good, but unfortunately, despite decades of vaccine development, not a single human DNA vaccine has been approved (Porter and Raviprakash 2017). This may sound surprising, even though this type is easily and conveniently designed on a computer, assembled using genetic engineering methods, and then manufactured using proven **biotechnology** techniques.

A look at the clinical data shows that there are still weaknesses to be resolved before **DNA vaccines** are available in humans for various infectious diseases. Many clinical trials have been unsuccessful predominantly due to a lack of **immunogenicity** or **efficacy** (Liu 2019), meaning that the immune response was either absent, too weak, and transient. An important reason is the instability of DNA after administration. Much of the DNA is degraded in tissues by certain enzymes called

**nucleases** before they can be taken up by cells and exert their effects. These enzymes are found in large quantities in our tissues. Therefore, the time between injection of the vaccine and uptake by the cells is particularly critical, almost a race against time. In addition, some cells are reluctant to take up larger **DNA molecules.** The fewer plasmids are taken up into the cells, the less protein is produced, and the less protein is produced, the weaker the immune response. Other approaches try to make the DNA more stable or to speed up or improve uptake. Overall, these efforts have already led to better immunogenicity (Suschak et al. 2017). One example is injection into the skin followed by **electroporation.** Here, an electric field briefly makes the cell membrane of the cells more permeable to biomolecules, including DNA (Li and Petrovsky 2016).

With **DNA vaccines,** there is a theoretical risk that the DNA will permanently insert into the genome of the recipient, although the risk has been found to be very low (Li and Petrovsky 2016). Plasmids must enter the nucleus in order for the genetic information to be translated into mRNA, and this proximity opens up the possibility of certain regions of the plasmid integrating into the genome of the body cell. Regarding a DNA vaccine for fish, the Norwegian Medicines Agency has answered no to this question, meaning that the vaccinated fish are not considered **genetically modified organisms** (GMOs). In this case, this means that the risk of the DNA remaining permanently or being stored was not considered relevant.

Although **DNA technology** is a phenomenal technique that has the potential, along with other technologies, to shape the vaccine landscape of the twenty-first century, it was assessable at the outset that it would not yield the first vaccines against **COVID-19.** Nevertheless, progress has been surprisingly good. There are many candidates in clinical phases, including the latest phase III clinical trial (WHO 2020a). At the forefront is Inovio Pharmaceuticals' vaccine candidate **INO-4800,** whose Phase I results were very good (Tebas et al. 2021) and has now progressed further (WHO 2020a). Although success cannot yet be estimated, it is to be hoped that this technology will also finally achieve the long-awaited breakthrough in COVID-19.

## 5.2 mRNA Vaccines

**RNA vaccines** are the newest of the technologies described here. The idea behind this is to use not DNA but another nucleic acid, the messenger ribonucleic acid **(mRNA).** This is the intermediate stage in gene expression from DNA, via RNA to protein.

In this technology, as in the other technologies, important proteins or protein parts of **SARS-CoV-2** are first identified. The genetic sequence for this protein is then inserted into a **DNA plasmid** using genetic engineering methods. Unlike DNA vaccines, however, the plasmid is not propagated and harvested in bacteria, for example. During production, large amounts of **mRNA** are produced with the help of certain enzymes, so-called RNA polymerases, using the genetic information of the DNA plasmid. This production takes place in vitro, i.e. outside living organisms, in a test tube. Subsequently, the mRNA is purified (i.e. DNA, enzymes and other components are removed) and additionally modified to make it more stable (Schlake et al. 2012). This mRNA is injected into the tissue (e.g. under the skin or into the muscle) as a vaccine. The mRNA is taken up by cells and the protein is formed based on the genetic information provided. Unlike **DNA vaccines,** which not only have to enter the cell but also the cell nucleus, entry into the cytoplasm of the cell is sufficient for mRNA, since proteins are formed here on the basis of the genetic information of mRNA (Fig. 5.2).

RNA is considered to be even more unstable than DNA and is also quickly cut by certain enzymes, so-called **ribonucleases.** For this reason, this technology was initially unattractive to the pharmaceutical industry for a long time (Kowalski et al. 2019). Nowadays, due to technical advances, there is the possibility to specifically improve the stability of RNA, e.g. by chemical modification. There are now two types (Pardi et al. 2018; Kowalski et al. 2019):

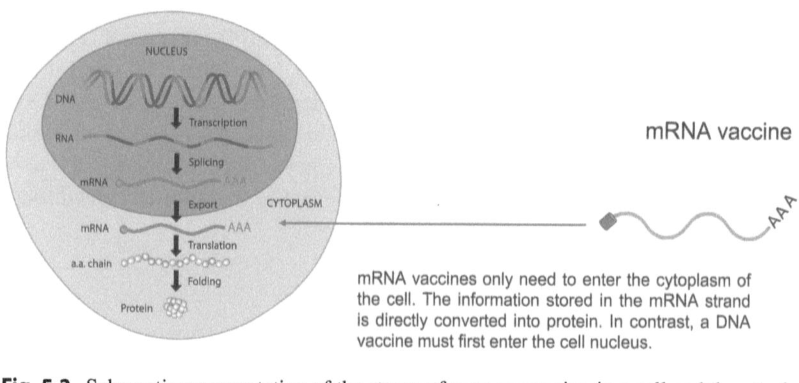


**Fig. 5.2** Schematic representation of the stages of gene expression in a cell and the attachment site of mRNA vaccines. (Source: Created using and modification of Adobe Stock, File No.: 166185134)

- **Non-replicating RNA:** This type is followed in most research projects, including the development of a vaccine against COVID-19, and is based on the above description.
- **Self-replicating RNA:** Here, in addition to the sequence for the viral protein, the RNA also contains sequences for replication enzymes. These enzymes are formed after introduction into the body cell and multiply the mRNA, just as a virus would do, which smuggles its RNA into the cell and then produces many copies of mRNA so that sufficient proteins can be formed.

One advantage of **mRNA vaccines** is the speed, flexibility and adaptability of the technology. This could produce vaccine candidates against new infectious diseases particularly quickly in the future. This type is considered to be very safe. Another advantage of mRNA vaccines over DNA vaccines is that there is no risk of the nucleic acid integrating into the genome of the vaccinated person, as it is not transported into the nucleus of our body cells (Schlake et al. 2012). A disadvantage could be the short residence time in the cell. The half-life of mRNA in cells is comparatively short, ranging from a few minutes to hours to a few days.

**mRNA vaccine candidates** are considered to be **highly immunogenic.** Numerous clinical trials are already underway for use in a wide variety of infectious diseases (Kowalski et al. 2019). RNA itself has immunostimulatory effects. However, one does not wish the immune response to be directed in full force against the mRNA that is injected into the tissue, but mainly against the viral proteins that are subsequently formed. For this reason, the activation of the immune system must be finely balanced. If the RNA product has too strong an immunostimulatory effect, the immune system does not wait long at all for proteins to be formed; instead, the mRNA is attacked directly. This then leads to the formation of a smaller amount of viral protein, which reduces the effect of the vaccine. On the other hand, the RNA must not be too "harmless" that no reaction occurs at all. There are various approaches, so-called **mRNA technology platforms**, to achieve this (Pardi et al. 2018). One example is the administration of the mRNA together with "dummy" RNA. This second RNA is only intended to activate the immune system and therefore has an amplification effect **(adjuvant effect).**

One of the most important aspects of **mRNA vaccines** is to get the molecules into somatic cells quickly and safely, much like DNA vaccines. One approach is **lipid nanoparticles** that form small vesicles and both protect and transport the mRNA into somatic cells (Reichmuth et al. 2016; Kowalski et al. 2019). However, the optimal form of production is still the subject of intensive research.

There were high hopes on the RNA vaccines with regard to **COVID-19** from the beginning. I myself was very skeptical, as the prospect of having vaccines soon was raised many years ago (Schlake et al. 2012), and by the end of 2019, no candidate had the prospect of gaining approval in the next few years. Moreover, just before the COVID-19 pandemic, there were sometimes mixed immunogenicity data in studies, which is why early researchers feared that RNA vaccines were facing a fate similar to DNA vaccines (Liu 2019). This makes the rapid success of this technology all the more remarkable. The first two candidates, BioNTech/Pfizer's **Comirnaty** (BNT162b2) and Moderna's **mRNA-1273**, both produced impressive efficacy data almost simultaneously (almost inconceivably). For example, good **tolerability, safety** and **immunogenicity** data from various clinical trials were already published for Comirnaty in Q3 2020 (Sahin et al. 2020). This was shortly followed by 95% efficacy data in December. Further, 2-month subject observation revealed a **safety profile** comparable to other vaccines (Polack et al. 2020). In principle, vaccines are somewhat more difficult to handle logistically than vector vaccines, as they must be stored at −70 °C (Comirnaty) or −20 °C (mRNA-1273). In this case, this difficulty was compensated by the establishment of vaccination centres. Another impressive aspect is the previously unanticipated high efficacy in elderly individuals (Pawelec and McElhaney 2021). This is also evident beyond clinical trials, for example in Israel, where high vaccination coverage has already been achieved. Here, a high protective effect was also observed after just one dose of vaccination (Hunter and Brainard 2021 preprint). These two candidates will continue to contribute significantly to bringing the **COVID-19 pandemic** under control through vaccination.

So far, no start-up difficulties in the **pharmaceutical process** seem to have significantly affected the ability to deliver. This was not readily expected, given the lack of experience with routine production. This is proof of how efficiently the preconditions for mass production have been created in the meantime, in parallel with the clinical trials. Chapeau! In addition, further candidates are in the pipeline, such as the candidate **CvnCoV** from the German company CureVac (WHO 2020a).

# 6 Other Aspects: Immunity and Viral Mutations

## 6.1 Vaccination Success in Old Age, Sterilising Immunity, Herd Immunity and Duration of Immunity

In the case of **COVID-19**, elderly people and people with certain pre-existing conditions are among the **high-risk groups,** in whom an infection can lead to particularly severe courses and even death. However, vaccinations are a particular challenge for older people, as the activity of the immune system decreases with age. This effect is called **immune senescence.** A summary review of various studies on the **efficacy** of influenza vaccines, for example, found 17–53% efficacy in older people, compared to 70–90% in younger people (Goodwin et al. 2006). Science has been trying for years to develop ideas and approaches to improve the efficacy of influenza vaccines specifically in the elderly (Smetana et al. 2018). For this reason, there are constant technological improvements, such as the use of better adjuvants, to improve the immune response. Nevertheless, the numbers vary depending on the vaccine, region and period. An assessment of the effectiveness of influenza vaccination since 2005 also found moderate rates, up to an average of 60% (Bouvier 2018). Nevertheless, it is of course important to get vaccinated against influenza regularly.

Due to this situation, it was initially feared that the **efficacy of COVID-19** would be significantly worse in the elderly. But, on the contrary, the results from the clinical trials and since the beginning of the vaccination campaign regarding the efficacy in the elderly especially with the mRNA vaccines were not impressive, but sensational. Personally, I never expected such a resounding success of the first

© The Author(s), under exclusive license to Springer Fachmedien Wiesbaden GmbH, part of Springer Nature 2023

P. U. B. Vogel, *COVID-19: Search for a vaccine*, essentials,
https://doi.org/10.1007/978-3-658-38931-4_6

generation vaccines in the elderly, as the pharmaceutical companies hardly had time to develop optimized formulations specifically for the elderly.

Precisely this success also has special relevance for dealing with another aspect of immunity, namely the question of whether immunity protects against infection or only prevents serious illness. Immunity that blocks the pathogen, i.e. that prevents the body from becoming infected at all, is called **sterilizing immunity.** In non-sterilizing immunity, vaccination prevents disease, but does not prevent the pathogen from infecting cells and reproducing in the vaccinated person for a period of time if contact occurs again. I had addressed the concept of sterilizing immunity in the first edition, as this was completely absent from public debates and there were also interim considerations in Germany towards the middle of 2020 to issue a health passport to recovering or vaccinated persons, i.e. to classify them as immune and harmless. In a worst-case scenario, this could have led to repeated unrecognised entries of the virus into, for example, senior citizens' facilities and thus to outbreaks. However, since the **high-risk groups** can now be effectively protected against **COVID-19**, the significance of this difference is significantly weakened.

Furthermore, the vaccination campaign launched against **COVID-19** aims to achieve herd immunity, as this can prevent the severe spread of a disease. The hope is that this will eliminate the need for harsh measures. However, whether or not vaccination induces **sterilizing immunity** does not determine whether **herd immunity** can be achieved with vaccination campaigns; in either case, achievement is possible. In the case of COVID-19, immunity could be non-sterilizing. This is suggested by previous infection trials with one of the common **cold coronaviruses, 229E.** Here, human volunteers were infected with this virus via the nose. Most developed cold symptoms and antibodies. Antibody titers dropped significantly within the first 12 months, but subjects were protected against re-infection 12 months after the first. None of the subjects developed cold symptoms, but the subjects did become infected, and excreted the virus for several days (Callow et al. 1990). Although not generalizable 1:1 to **SARS-CoV-2** due to the highly adapted nature of the type used, which has circulated in human populations for hundreds of years (Graham et al. 2013), this would not be atypical. The effect of non-sterilizing immunity is commonly observed in respiratory infectious diseases, such as influenza (Bouvier 2018), but also in bacterial diseases such as pertussis (Solans and Locht 2019). Regardless of which species confers immunity, herd immunity is possible in both cases, except that depending on the expression of other aspects, control of an infectious disease may be somewhat more difficult if non-sterilizing immunity exists. For example, pertussis was very well controlled by blanket vaccination. After some countries switched to a safer vaccine, but unfortunately one that confers short-lived immunity, there have been cyclical epidemics of pertussis at intervals of a few years because the pathogen is already circulating subclinically in some people and then can quickly cause disease again as immunity wanes in the population. However, dimensions are also important. Non-sterilizing immunity does not mean that the virus circulates in vaccinated populations in the same way as in non-vaccinated populations. Experiments with coronavirus 229E already showed that subjects excreted the virus for a significantly shorter time after 12 months (Callow et al. 1990). Therefore, the overall burden of virus circulation is significantly reduced in vaccinated populations.

Interestingly, in the case of **COVID-19,** initial studies with the available vaccines indicate that infections can also be strongly prevented. This may depend on the vaccine used and must also be shown over a longer period of time. Certain antibodies, so-called **IgA antibodies,** are often responsible for sterilising immunity. These also have the ability to stop infections in influenza, for example, depending on the immunization in animal models (Bouvier 2018). However, these IgA antibodies often break down within a few months and are not as long-lasting as, for example, the IgG antibodies found in blood and tissue fluids. For this reason, it will be important to show that suppression of subclinical infections is also long-lasting, although, as mentioned above, this issue has become somewhat less important due to the high protective effect of **COVID-19 vaccination** in the elderly, although, of course, increased subclinical circulation could also promote **viral mutations**.

Some aspects of COVID-19, such as the **duration of immunity,** are still unclear. In the case of **SARS,** a possible immunity of 3 years was estimated based on analysis of the antibody history of SARS patients (Wu et al. 2007), although it is still unclear whether antibodies alone are sufficient to assess immunity. In addition, infection trials with **229E** at the time demonstrated effective protection of at least one year, although later time points have not been verified (Callow et al. 1990). Also, the fact that common corona cold viruses cause epidemic clustering at 2–3 year intervals in some regions (Greenberg 2016) may suggest 2–3 year immunity, although other causes are also possible. More data are needed for **COVID-19,** but current estimates are positive following natural infections. Based on various immunological markers (antibodies, T cells), COVID-19 is inferred to have an immune protection of already about 8 months (Reynolds 2021). However, it must be kept in mind that immunity after vaccination may resemble natural infection, but may also be shorter. This depends on various factors. Therefore, it will be very important to follow up the data from subjects of clinical trials. One positive aspect is the timing of clinical trials. Since the extensive vaccinations in clinical trials have already taken place from mid-2020 onwards, we will be in a position to make a good estimate of the provisional duration of immunity in the summer of 2021. Insofar as immunity should fade, there would still be enough time to start with

**booster vaccinations** before the next cold spell without running into logistical problems, since by late summer everyone who is vaccine-ready will probably have received their first vaccination, at least in Germany. However, the question of whether soft measures are sufficient depends on more than just the factors mentioned here. **Virus mutations** also have an influence on whether soft measures will be sufficient once herd immunity has been achieved. With regard to possible booster vaccinations, it still needs to be clarified whether different, successive vaccines in combination achieve the same effect or not. Especially in the case of vector vaccines, with regard to booster vaccinations, it must be examined whether this leads to reduced efficacy or whether the vaccines can be used repeatedly, which would be desirable. Thus, many more scientific studies and time are needed to analyze in more detail the duration of immunity after **COVID-19 vaccination** and other important aspects of **SARS-CoV-2.**

## 6.2 Virus Mutations

A current topic that has gained increasing attention towards the end of 2020 are **mutations** of the pathogen **SARS-CoV-2.** These virus variants are often commonly called British, Brazilian or South African variants after the place of their discovery. In addition to spreading more rapidly, there are some fears that these or new variants could undermine the effectiveness of existing vaccines.

First, an important point. **Viral mutations** occur in every infected person. A common infection with **SARS-CoV-2** involves the infection of up to millions of the body's own cells (Sender et al. 2021 preprint). Upon infection, the infected cells are reprogrammed into virus factories that subsequently release large numbers of new virus particles. It is estimated based on other coronaviruses that approximately 100 new infectious virus particles are generated per virus-infected cell (Bar-On et al. 2020). However, many errors occur during the replication of viruses. For example, incorrect nucleotides are inserted into the viral genome, but in some cases a nucleotide or short pieces of sequence are also lost (= deletion) or additionally inserted (= insertion). **RNA viruses** generally have no control over the correctness of the replication of their viral genome, i.e. errors in the formation of the RNA strand are not corrected, which means that RNA viruses in particular have a very high mutation rate, even compared to DNA viruses (Sanjuán et al. 2010). Coronaviruses are again unique in that they have an enzymatic activity (**ExoN**) that corrects errors in specific regions of the **viral genome** (de Witt et al. 2016). Coronaviruses are estimated to have a substitution rate (replacement of nucleotides by others) of $10^{-4}$ substitutions per position per year (Ye et al. 2020).

The **mutations** in the virus genome are either lethal for the virus (virus can no longer reproduce), neutral (no significant change) or positive (the virus acquires a new beneficial property). Most mutations are lethal, i.e. certain genes of the viral genome are defective. To the extent that a virus particle gains an advantage from a mutation, such as faster replication, it will prevail over the other original virus particles in the long run. Although **viral mutations** occur in every human being, these viral variants with new positive properties arise only very rarely.

In the case of **SARS**, for example, sequence analyses revealed that the virus isolates from patients at a later stage of the pandemic lacked a **29-nucleotide sequence** in the virus genome compared with the animals examined in the outbreak region. The exact significance remained unclear, but there were various conjectures, e.g. that this mutation either originated in animals and enabled the so-called **spillover,** i.e. transmission from animals to humans, or that this loss only occurred in humans, representing a kind of adaptation, as it were, which enabled more efficient human-to-human transmission (Kahn and McIntosh 2005). Ultimately, however, later analyses showed that the loss of this 29-nucleotide sequence led to the attenuation of the virus, i.e. it was a disadvantage for the spread of the virus (Muth et al. 2018).

The normal **cold coronaviruses** show little propensity to change their properties, although they do exhibit some genetic variability, as exemplified by a study examining the abundance and **genetic variability** of these coronaviruses in school children over a period of approximately 1.5 years (Liu et al. 2017). To this end, it is hypothesized that different variants in the receptor-binding domain of the **spike protein** have continuously replaced each other over the past 50 years (Wong et al. 2017). Nevertheless, their property of triggering only mild colds in healthy individuals has been stable over decades. For example, there is only one case in which a more aggressive viral isolate of coronavirus **NL63** was found. For this reason, it is assumed as a rule of thumb that coronaviruses become less pathogenic through increasing adaptation to humans (Ye et al. 2020). However, SARS-CoV-2 has been circulating in humans for a short time. For example, a viral variant was found that had a mutation in the spike protein that affected only one amino acid. This viral variant showed increased pathogenicity (Becerra-Flores and Cardozo 2020). This variant, D614G, was first discovered in April 2020 and has become the dominant variant worldwide (Hohmann-Jeddi 2021). A correlation between mutation occurrence and hard **lockdown** has also been found. After initially higher mutation rates, mutations stabilized to a few positions in some countries with hard lockdown (Pachetti et al. 2020), probably related to the suppressed circulation of the virus.

However, in the case of **mutations,** it is rather unusual for a more aggressive (virulent) isolate to prevail without acquiring any additional advantage, such as better transmission. Only **more pathogenic variants** would, on average, make people more likely to seek medical attention and lead to increased admissions to isolation or intensive care units where healthcare professionals protect themselves from infection by wearing special professional clothing, while on average less severely ill individuals (less virulent isolate) would be more likely to continue to participate in social life where the risk of transmission to others is greater. It is somewhat different for infectious diseases for which vaccines are used. For example, genetic alteration of a viral isolate could cause it to behave the same in non-vaccinated individuals before and after mutation, but undermine the immunity of vaccinated individuals. Then the isolate is not more pathogenic, but has nevertheless acquired an advantage. This is seen, for example, in **influenza viruses**, whose constant genetic change is called **antigenic drift** (Bouvier and Palese 2008). However, as mentioned above, this property is not a process previously known to be typical of coronaviruses. However, there is another mechanism of genetic change in coronaviruses that can spontaneously involve large changes. This mechanism is called **recombination** and refers to the exchange of genetic regions of the viral genome between two different coronaviruses that replicate in the same cell. This type of genetic change has been associated with the emergence of pathogenic variants in animals, e.g. the emergence of a coronavirus pathogenic to dogs by recombination with a porcine coronavirus (Ntafis et al. 2011). In the emergence of new **viral mutants,** the receptor-binding domain of the spike protein is an important domain. Using a recombinant **SARS-CoV-2** spike protein construct, it was shown that mutations in the receptor-binding domain, among others, can lead to escape variants that subvert the action of neutralizing antibodies (Weisblum et al. 2020).

The **variants** currently classified as being of concern are B.1.1.7, B.1.351 and P.1, as well as further subtypes thereof. The so-called **British variant B.1.1.7** has a higher transmissibility (RKI 2021). This variant carries a large number of mutations, including 8 in the spike protein, which is important for cell infection (Hohmann-Jeddi 2021). For this reason, there have been several warnings in Germany against the rapid spread of this variant and thus, among other things, the harsh **lockdown measures** have been defended. In the meantime, this variant also dominates in Germany. However, it looks as if it will not completely undermine the current vaccines, as severe courses of the disease can be prevented with the mRNA vaccines. The first vector vaccine licensed in Germany also appears to protect effectively against this variant (Bäuerle 2021). However, a subform has emerged from B.1.1.7 that is more poorly neutralized by antibodies in vitro. The **South African variant B.1.351** is also spreading more rapidly and there is evidence that immunity or vaccination does not protect against this variant. The **Brazilian vari-**

**ant P.1** may also have advantages in transmission as well as the ability to subvert neutralization by existing antibodies (RKI 2021).

It is hard to predict what will happen in the next few months, which means that **virus mutations** could theoretically emerge that undermine the effectiveness of the first vaccines. In the long term, I suspect that we will not have to worry about seeing new **variants** later on, year after year or several times a year, against which none of the licensed vaccines will be effective. There are several new vaccines on track to receive approval in Europe and around the world. The existing range of vaccines will thus gradually be expanded, including completely different types and based on different antigenic structures (over 200 projects in the pipeline). This will make it increasingly difficult for the virus to form **escape mutants** against which no vaccine is effective. There are also limits to viruses, an infinite variation of the **spike protein** is rather unlikely, as in many cases it would no longer fit the receptor.

# Summary and Outlook 7

The **COVID 19 pandemic** is unprecedented in scale and duration. The health, social and economic impacts were and still are significant. The harsh measures, including lockdowns in the spring of 2020 and the winter period of 2020/2021, were necessary to prevent **COVID-19** from becoming probably the second deadliest event in human history after the Spanish flu (Vogel and Schaub 2021). Nevertheless, despite government support, it is not possible for many companies in diverse industries to survive the harsh measures in the long run without immense permanent damage. For this reason, the use of vaccines is the most important preventive means of returning to whatever normality is possible. The achievement of science and the pharmaceutical industry, supported by governments, institutions and the regulatory authorities, to develop a vaccine in less than 12 months, to test it for suitability in large-scale clinical trials, to get it approved, to go into mass production and to deliver it, is really only one thing – sensational! Even if COVID-19 will go down in history as one of the worst **pandemics**, this unprecedented success story of vaccine development will not be forgotten. I would not be surprised if this were subsequently honoured with a **Nobel Prize**.

The current figures, including the first **vaccine approvals** and the increasing vaccination rate, but also the prospect of further approvals of new vaccines, are very encouraging. The sharp drop in the number of infections in countries that are particularly far advanced with vaccinations, such as the USA, the UK or Israel, shows the effect that these vaccination campaigns are having. Add to that the summer of 2021, when cold infection numbers usually decline. In combination, a likely scenario is that the harsh measures will be gradually reduced over the course of the spring or early summer, and that we will be able to get by with at most soft measures such as spacing, mouth-to-nose coverage, hygiene, and the targeted use of

© The Author(s), under exclusive license to Springer Fachmedien Wiesbaden GmbH, part of Springer Nature 2023
P. U. B. Vogel, *COVID-19: Search for a vaccine*, essentials,
https://doi.org/10.1007/978-3-658-38931-4_7

diagnostic tests during the next cold snap in 2021/2022. However, the outcome will also be influenced by possible new **virus variants**, whose emergence and ability to possibly undermine the effectiveness of existing vaccines cannot be predicted. It will also be important to demonstrate the **safety** of the vaccines for children, as the use of soft measures will also require kindergartens and schools to be open.

There are various factors that can influence the achievement of the goal of "normality". The **vaccine supply** plays a major role in this. The quantities of vaccine that have been promised or assured until autumn 2021 will be sufficient to offer vaccination to all people willing to be vaccinated, at least in Germany. However, there are some uncertainties. Shortly after approval, some manufacturers experienced temporary production line stoppages. A failure of promised vaccine deliveries can have various reasons, ranging from plant breakdowns, shortages of required starting materials, exceeding of permissible limits in campaign production, unexpectedly low yields of intermediates, and many others are possible. Temporary inability to supply also occurs from time to time with other vaccines and would not be unusual, especially since production is running at full speed. Currently, production capacity continues to increase, which is very important. In addition, it will be very important to expand the existing supply through further **vaccine approvals** in order to avoid or compensate for any supply bottlenecks and, against the background of new **virus variants,** to have a broad vaccine arsenal available. However, the **duration of immunity** is also important and remains to be seen.

Of course, the **willingness** of the population to be vaccinated also plays an important role, which according to surveys is moving in the right direction in Germany. With every month that the worldwide number of vaccinated people increases without significant **side effects** attributable to the vaccine becoming known, confidence in vaccines will continue to increase. This is precisely the point that is already readily assessable, as nearly 400 million people worldwide have received a vaccination without any rates of adverse reactions deemed unusual by authorities. However, temporary stops of individual vaccines or vaccine batches may also occur in the future in order to examine certain safety aspects more closely.

# What You Can Take Away from this *Essential*

- New viruses with high health relevance represent an enormous challenge for health systems and the pharmaceutical industry.
- Newer technologies offer the possibility of shortening vaccine development times and have made a breakthrough with COVID-19.
- Rapid approval of vaccines is necessary in the event of pandemics, but it also entails risks.
- An outstanding aspect is the suitability of the vaccines for high-risk groups.

© The Author(s), under exclusive license to Springer Fachmedien Wiesbaden GmbH, part of Springer Nature 2023
P. U. B. Vogel, *COVID-19: Search for a vaccine*, essentials,
https://doi.org/10.1007/978-3-658-38931-4

# References

Aps LRMM, Piantola MAF, Pereira SA et al (2018) Adverse events of vaccines and the consequences of non-vaccination: a critical review. Rev Saude Publica 52:40. https://doi.org/10.11606/s1518-8787.2018052000384

ärzteblatt.de (2021) Russland kündigt Massenproduktion von zweitem Impfstoff an. https://www.aerzteblatt.de/nachrichten/120547/Russland-kuendigt-Massenproduktion-von-zweitem-Impfstoff-an. Accessed 25 February 2021

Awadasseid A, Wu Y, Tanaka Y et al (2021) Current advances in the development of SARS-CoV-2 vaccines. Int J Biol Sci 17:8–19. https://doi.org/10.7150/ijbs.52569

Balakrishnan VS (2020) The arrival of Sputnik V. Lancet Infect Dis 20:1128. https://doi.org/10.1016/S1473-3099(20)30709-X

Balfour H (2021) Valneva may provide Europe with the only inactivated virus vaccine for COVID-19. https://www.europeanpharmaceuticalreview.com/news/139688/valneva-may-provide-europe-with-the-only-inactivated-virus-vaccine-for-covid-19/. Accessed 25 February 2021

Banerjee A, Kulcsar K, Misra V et al (2019) Bats and coronaviruses. Viruses 11:41. https://doi.org/10.3390/v11010041

Bar-On YM, Flamholz A, Phillips R et al (2020) SARS-CoV-2 (COVID-19) by the numbers. Elife 9:e57309. https://doi.org/10.7554/eLife.57309

Bäuerle A (2021) Analyse des RKI Coronavirus-Varianten breiten sich aus. ÄrzteZeitung. https://www.aerztezeitung.de/Nachrichten/Coronavirus-Varianten-breiten-sich-weiter-aus-416929.html. Accessed 15 February 2021

Becerra-Flores M, Cardozo T (2020) SARS-CoV-2 viral spike G614 mutation exhibits higher case fatality rate. Int J Clin Pract 74:e13525. https://doi.org/10.1111/ijcp.13525

Bijlenga G, Cook JKA, Gelb J Jr et al (2004) Development and use of the H strain of avian infectious bronchitis virus from the Netherlands as a vaccine: a review. Avian Pathol 33:550–557. https://doi.org/10.1080/03079450400013154

Bouvier NM (2018) The future of influenza vaccines: a historical and clinical perspective. Vaccines 6:58. https://doi.org/10.3390/vaccines6030058

© The Author(s), under exclusive license to Springer Fachmedien Wiesbaden GmbH, part of Springer Nature 2023
P. U. B. Vogel, *COVID-19: Search for a vaccine*, essentials,
https://doi.org/10.1007/978-3-658-38931-4

Bouvier NM, Palese P (2008) The biology of influenza viruses. Vaccine 4:D49–D53. https://doi.org/10.1016/j.vaccine.2008.07.039

Callow KA, Parry HF, Sergeant M, Tyrrell DA (1990) The time course of the immune response to experimental coronavirus infection of man. Epidemiol Infect 105:435–446. https://doi.org/10.1017/s0950268800048019d

CDC (2014) Global health security: immunization. https://www.cdc.gov/globalhealth/security/immunization.htm. Accessed 20 February 2021

CDC (2021) Allergic reactions including anaphylaxis after receipt of the first dose of Pfizer-BioNTech COVID-19 vaccine – Unites States, December 14–23, 2020. MMWR Morb Mortal Wkly Rep 70:46–51. https://doi.org/10.15585/mmwr.mm7002e1

Chan-Yeung M, Xu RH (2003) SARS: epidemiology. Respirology 8:9–14. https://doi.org/10.1046/j.1440-1843.2003.00518.x

Chroboczek J, Szurgot I, Szolajska E (2014) Virus-like particles as vaccine. Acta Biochim Pol 61:531–539

Cision PR Newswire (2020) Codagenix and Serum Institute of India Initiate Dosing in Phase 1 Trial of COVI-VAC, a Single Dose, Intranasal, Live Attenuated Vaccine for COVID-19. https://www.prnewswire.com/news-releases/codagenix-and-serum-institute-of-india-initiate-dosing-in-phase-1-trial-of-covi-vac-a-single-dose-intranasal-live-attenuated-vaccine-for-covid-19-301203130.html. Accessed 21 January 2021

Corman VM, Muth D, Niemeyer D et al (2018) Hosts and sources of endemic human coronaviruses. Adv Virus Res 100:163–188. https://doi.org/10.1016/bs.aivir.2018.01.001

Crisci E, Bárcena J, Montoya M (2013) Virus-like particle-based vaccines for animal viral infections. Immunologia 32:102–116. https://doi.org/10.1016/j.inmuno.2012.08.002

CSSE (2021) Coronavirus 2019-nCoV global cases by Johns Hopkins CSSE. https://gisanddata.maps.arcgis.com/apps/opsdashboard/index.html#/bda7594740fd40299423467b48e9ecf6. Accessed 18 März 2021

D'alò GL, Zorzoli E, Capanna A et al (2017) Frequently asked questions on seven rare adverse events following immunization. J Prev Med Hyg 58:E13–E26

de Wit E, van Doremalen N, Falzarano D et al (2016) SARS and MERS: recent insights into emerging coronaviruses. Nat Rev Microbiol 14:523–534. https://doi.org/10.1038/nrmicro.2016.81

Dittmann S (2002) Risiko des Impfens und das noch größere Risiko, nicht geimpft zu sein. Bundesgesundheitsbl – Gesundheitsforsch – Gesundheitsschutz 45:316–322. Springer. https://www.rki.de/DE/Content/Infekt/Impfen/Bedeutung/Downloads/Dittmann_Risiko.pdf?__blob=publicationFile. Accessed 22 February 2021

Drosten C, Günther S, Preiser W et al (2003) Identification of a novel coronavirus in patients with severe acute respiratory syndrome. N Engl J Med 348:1967–1976. https://doi.org/10.1056/NEJMoa030747

Fehr AR, Perlman S (2015) Coronaviruses: an overview of their replication and pathogenesis. Methods Mol Biol 1282:1–23. https://doi.org/10.1007/978-1-4939-2438-7_1

Felberbaum RS (2015) The baculovirus expression vector system: a commercial manufacturing platform for viral vaccines and gene therapy vectors. Biotechnol J 10:702–714. https://doi.org/10.1002/biot.201400438

Gaglia MM, Munger K (2018) More than just oncogenes: mechanims of tumorigenesis by human viruses. Curr Opin Virol 32:48–59. https://doi.org/10.1016/j.coviro.2018.09.003

Gao Q, Bao L, Mao H et al (2020) Rapid development of an inactivated vaccine candidate for SARS-CoV-2. Science 6:eabc1932. https://doi.org/10.1126/science.abc1932
Garg H, Mehmetoglu-Gurbuz T, Joshi A (2020) Virus like particles (VLP) as multivalent vaccine candidate against Chikungunya, Japanese Encephalitis, Yellow Fever and Zika virus. Sci Rep 10:4017. https://doi.org/10.1038/s41598-020-61103-1
Gerdts V, Zakhartchouk A (2017) Vaccines for porcine epidemic diarrhea virus and other swine coronaviruses. Vet Microbiol 206:45–51. https://doi.org/10.1016/j.vetmic.2016.11.029
Goodwin K, Viboud C, Simonsen L (2006) Antibody response to influenza vaccination in the elderly: a quantitative review. Vaccine 24:1159–1169. https://doi.org/10.1016/j.vaccine.2005.08.105
Graham RL, Donaldson EF, Baric RS (2013) A decade after SARS: strategies for controlling emerging coronaviruses. Nat Rev Microbiol 11:836–848. https://doi.org/10.1038/nrmicro3143
Greenberg SB (2016) Update on human rhinovirus and coronavirus infections. Semin Respir Crit Care Med 37:555–571. https://doi.org/10.1055/s-0036-1584797
Halstead SB, Katzelnick L (2020) COVID-19 vaccines: should we fear ADE? J Infect Dis 222:1946–1950. https://doi.org/10.1093/infdis/jiaa518
Hamilton K, Visser D, Evans B et al (2015) Identifying and reducing remaining stocks of rinderpest virus. Emerg Infect Dis 21:2117–2121. https://doi.org/10.3201/eid2112.150227
Hampton LM, Aggarwal R, Evans SJW et al (2021) General determination of causation between COVID-19 vaccines and possible adverse events. Vaccine 39:1478–1480. https://doi.org/10.1016/j.vaccine.2021.01.057
Heikkinen T, Järvinen A (2003) The common cold. Lancet 361:51–59. https://doi.org/10.1016/S0140-6736(03)12162-9
Heim A (2016) Adenoviren. In: Suerbaum S et al (eds) Medizinische Mikrobiologie und Infektiologie. Springer, Heidelberg. https://doi.org/10.1007/978-3-662-48678-8_70
Hohmann-Jeddi C (2021) SARS-Coronavirus-2 Virusvarianten im Überblick. Pharmazeutische Zeitung online. 20.02.2021. https://www.pharmazeutische-zeitung.de/virusvarianten-im-ueberblick-123903/. Accessed 22 February 2021
Hunter PR, Brainard J (2021) Estimating the effectiveness of the Pfizer COVID-19 BNT162b2 vaccine after a single dose. A reanalysis of a study of 'real-world' vaccination outcomes from Israel. MedRxiv XY. https://doi.org/10.1101/2021.02.01.21250957
Kahn JS, McIntosh K (2005) History and recent advances in coronavirus discovery. Pediatr Infect Dis J 24:223–227. https://doi.org/10.1097/01.inf.0000188166.17324.60
Karch CP, Burkhard P (2016) Vaccine technologies: from whole organisms to rationally designed protein assemblies. Biochem Pharmocol 120:1–14. https://doi.org/10.1016/j.bcp.2016.05.001
Kim JHK, Marks F, Clemens JD (2021) Looking beyond COVID-19 vaccine phase 3 trials. Nat Med 27:205–211. https://doi.org/10.1038/s41591-021-01230-y
Knoll MD, Wonodi C (2021) Oxford-AstraZeneca COVID-19 vaccine efficacy. Lancet 397:72–74. https://doi.org/10.1016/S0140-6736(20)32623-4
Kowalski PS, Rudra A, Miao L et al (2019) Delivering the messenger: advances in technologies for therapeutic mRNA delivery. Mol Ther 27:710–728. https://doi.org/10.1016/j.ymthe.2019.02.012

Kremer EJ (2020) Pros and cons of adenovirus-based SARS-CoV-2 vaccines. Mol Ther 28:2303–2304. https://doi.org/10.1016/j.ymthe.2020.10.002

Lambert PH, Ambrosino DM, Andersen SR et al (2020) Consensus summary report for CEPI/BC March 12–13, 2020 meeting: assessment of risk of disease enhancement with COVID-19 vaccines. Vaccine 38:4783–4791. https://doi.org/10.1016/j.vaccine.2020.05.064

Le Nouën C, Collins PL, Buchholz UJ (2019) Attenuation of human respiratory viruses by synonymous genome recoding. Front Immunol 10:1250. https://doi.org/10.3389/fimmu.2019.01250

Li L, Petrovsky N (2016) Molecular mechanisms for enhanced DNA vaccine immunogenicity. Expert Rev Vaccines 15d:313–329. https://doi.org/10.1586/14760584.2016.1124762

Li X, Zai J, Wang X, Li Y (2020) Potential of large "first generation" human-to-human transmission of 2019-nCoV. J Med Virol 92:448–454. https://doi.org/10.1002/jmv.25693

Liu MA (2019) A comparison of plasmid DNA and mRNA as vaccine technologies. Vaccines (Basel) 7:37. https://doi.org/10.3390/vaccines7020037

Liu P, Shi L, Zhang W, He J, Liu C et al (2017) Prevalence and genetic diversity analysis of human coronaviruses among cross-border children. Virol J 14:230. https://doi.org/10.1186/s12985-017-0896-0

Logunov DY, Dolzhikova IV, Shcheblyakov DV et al (2021) Safety and efficacy of an rAd26 and rAd5 vector-based heterologous prime-boost COVID-19 vaccine: an interim analysis of a randomised controlled phase 3 trial in Russia. Lancet 397:671–681. https://doi.org/10.1016/S0140-6736(21)00234-8

Lundstrom K (2019) RNA viruses as tools in gene therapy and vaccine development. Genes (Basel) 10:189. https://doi.org/10.3390/genes10030189

Masters PS (2006) The molecular biology of coronaviruses. Adv Virus Res 66:193–292. https://doi.org/10.1016/S0065-3527(06)66005-3

Meng J, Lee S, Hotard AL et al (2014) Refining the balance of attenuation and immunogenicity of respiratory syncytial virus by targeted codon deoptimization of virulence genes. mBio 5:e01704–e01714. https://doi.org/10.1128/mBio.01704-14

Mennechet FJD, Paris O, Ouoba AR et al (2019) A review of 65 years of human adenovirus seroprevalence. Expert Rev Vaccines 18:597–613. https://doi.org/10.1080/14760584.2019.1588113

Mercado NB, Zahn R, Wegmann F et al (2020) Single-shot Ad26 vaccine protects against SARS-CoV-2 in rhesus macaques. Nature 586:583–588. https://doi.org/10.1038/s41586-020-2607-z

Minor PD (2015) Live attenuated vaccines: historical successes and current challenges. Virology 479–480:379–392. https://doi.org/10.1016/j.virol.2015.03.032

Mohsen MO, Zha L, Cabral-Miranda G et al (2017) Major findings and recent advances in virus-like particle (VLP)-based vaccines. Semin Immunol 34:123–132. https://doi.org/10.1016/j.smim.2017.08.014

Morales-Sánchez A, Fuentes-Pananá EM (2014) Human viruses and cancer. Viruses 6:4047–4079. https://doi.org/10.3390/v6104047

Müller U, Vogel P, Alber G et al (2008) The innate immune system of mammals and insects. In: Egesten A, Schmidt A, Herwald H (eds) Contributions to Microbiology, vol 15. Karger, Basel, pp 21–44. https://doi.org/10.1159/000135684

Muth D, Corman VM, Roth H et al (2018) Attenuation of replication by a 29 nucleotide deletion in SARS-coronavirus acquired during the early stages of human-to-human transmission. Sci Rep 8:15177. https://doi.org/10.1038/s41598-018-33487-8

Ntafis V, Mari V, Decaro N, Papanastassopoulou M, Papaioannou N et al (2011) Isolation, tissue distribution and molecular characterization of two recombinant canine coronavirus strains. Vet Microbiol 151:238–244. https://doi.org/10.1016/j.vetmic.2011.03.008

OIE (2000) Chapter 2.3.2: Avian infectious bronchitis. https://www.oie.int/fileadmin/Home/eng/Health_standards/tahm/2.03.02_AIB.pdf. Accessed 30 June 2020

Pachetti M, Marini B, Giudici F et al (2020) Impact of lockdown on Covid-19 case fatality rate and viral mutations spread in 7 countries in Europe and North America. J Transl Med 18:338. https://doi.org/10.1186/s12967-020-02501-x

Pardi N, Hogan MJ, Porter FW et al (2018) mRNA vaccines – a new era in vaccinology. Nat Rev Drug Discov 17:261–279. https://doi.org/10.1038/nrd.2017.243

Pati R, Shevtsov M, Sonawane A (2018) Nanoparticle vaccines against infectious diseases. Front Immunol 9:2224. https://doi.org/10.3389/fimmu.2018.02224

Pawelec G, McElhaney J (2021) Unanticipated efficacy of SARS-CoV-2 vaccination in older adults. Immun Ageing 18:7. https://doi.org/10.1186/s12979-021-00219-y

PEI (2018) Press release: Modular virus-like particles as vaccine platform. https://www.pei.de/EN/newsroom/press-releases/year/2018/09-modular-virus-like-particles-as-vaccine-platform.html. Accessed 24 February 2021

PEI (2019) Weltweit erster Ebola-Impfstoff zugelassen. https://www.pei.de/DE/newsroom/hp-meldungen/2019/191113-erster-impfstoff-schutz-vor-ebola-zulassung-in-eu.html. Accessed 30 June 2020

PEI (2020) Influenza-Impfstoffe (alle Zulassungen). https://www.pei.de/DE/arzneimittel/impfstoffe/influenza-grippe/influenza-node.html. Accessed 30 June 2020

PEI (2021) Verdachtsfälle von Nebenwirkungen und Impfkomplikationen nach Impfung zum Schutz vor COVID-19. https://www.pei.de/SharedDocs/Downloads/DE/newsroom/dossiers/sicherheitsberichte/sicherheitsbericht-27-12-bis-12-02-21.pdf?__blob=publicationFile&v=7. Accessed 25 February 2021

Pfleiderer M, Wichmann O (2015) Von der Zulassung von Impfstoffen zur Empfehlung durch die Ständige Impfkommission in Deutschland. Bundesgesundheitsbl 08(01):2015. https://doi.org/10.1007/s00103-014-2109-y

Pharmazeutische Zeitung online (2020) Autoantikörper an COVID-19 Pathologie beteiligt. https://www.pharmazeutische-zeitung.de/weitere-infos/impressum/. Accessed 25 January 2021

Piedimonte G, Perez MK (2014) Respiratory syncytial virus infection and bronchiolitis. Pediatr Rev 35:519–530. https://doi.org/10.1542/pir.35-12-519

Polack FP, Thomas SJ, Kitchin N et al (2020) Safety and efficacy of the BNT162b2 mRNA Covid-19 vaccine. N Engl J Med 383:2603–2615. https://doi.org/10.1056/NEJMoa2034577

Porter KR, Raviprakash K (2017) DNA vaccine delivery and improved immunogenicity. Curr Issues Mol Biol 22:129–138. https://doi.org/10.21775/cimb.022.129

Prestel J, Volkers P, Mentzer D et al (2014) Risk of Guillain–Barré syndrome following pandemic influenza A(H1N1) 2009 vaccination in Germany. Pharmacoepidemiol Drug Saf 23:1192–1204

Ramakrishnan S, Kappala D (2019) Avian infectious bronchitis virus. In: Malik YS, Singh RK, Yadav MP (eds) Recent advances in animal virology, 1st edn. Springer, Singapore, pp 301–319. https://doi.org/10.1007/978-981-13-9073-9_16

Rauch S, Jasny E, Schmidt KE et al (2018) New vaccine technologies to combat outbreak situations. Front Immunol 9:1963. https://doi.org/10.3389/fimmu.2018.01963

Reichmuth AM, Oberli MA, Jaklenec A et al (2016) mRNA vaccine delivery using lipid nanoparticles. Ther Deliv 7:319–334. https://doi.org/10.4155/tde-2016-0006

Reynolds S (2021) Lasting immunity found after recovery from COVID-19. https://www.nih.gov/news-events/nih-research-matters/lasting-immunity-found-after-recovery--covid-19#:~:text=The%20immune%20systems%20of%20more,lasting%20immune%20memories%20after%20vaccination. Accessed 25 February 2021

Riedel S (2005) Edward Jenner and the history of smallpox and vaccination. Proc (Bayl Univ Med Cent) 18:21–25. https://doi.org/10.1080/08998280.2005.11928028

RKI (2021) Übersicht und Empfehlungen zu besorgniserregenden SARS-CoV-2 Virusvarianten (VOC). https://www.rki.de/DE/Content/InfAZ/N/Neuartiges_Coronavirus/Virusvariante.html. Accessed 26 February 2021

Rostad CA, Anderson EJ (2021) Optimism and caution for an inactivated COVID-19 vaccine. Lancet Infect Dis 21 S1473-3099(20):30988–30989. https://doi.org/10.1016/S1473-3099(20)30988-9

Sahin U, Muik A, Derhovanessian E et al (2020) COVID-19 vaccine BNT162b1 elicits human antibody and Th1 T cell responses. Nature 586:594–599. https://doi.org/10.1038/s41586-020-2814-7

Sanjuán R, Nebot MR, Chirico N et al (2010) Viral mutation rates. J Virol 84:9733–9748. https://doi.org/10.1128/JVI.00694-10

Saxena M, Van TTH, Baird FJ et al (2013) Pre-existing immunity against vaccine vectors – friend or foe? Microbiology 59:1–11. https://doi.org/10.1099/mic.0.049601-0

Schlake T, Thess A, Fotin-Mleczek M et al (2012) Developing mRNA-vaccine technologies. RNA Biol 9:1319–1330. https://doi.org/10.4161/rna.22269

Schriever J, Schwarz G, Steffen C et al (2009) Das Genehmigungsverfahren klinischer Prüfungen von Arzneimittel bei den Bundesoberbehörden. Bundesgesundheitsbl 52:377–386. https://doi.org/10.1007/s00103-009-0821-9

Sender R, Bon-On YM, Flamholz A et al (2020) The total number and mass of SARS-CoV-2 virions in an infected person. medRxiv. https://doi.org/10.1101/2020.11.16.20232009

Smetana J, Chlibek R, Shaw J et al (2018) Influenza vaccination in the elderly. Hum Vaccin Immunother 14:540–549. https://doi.org/10.1080/21645515.2017.1343226

Solans L, Locht C (2019) The role of mucosal immunity in pertussis. Front Immunol 9:3068. https://doi.org/10.3389/fimmu.2018.03068

Spencer JP, Pawlowski RHT, Thomas S (2017) Vaccine adverse events: separating myth from reality. Am Fam Physician 95:786–794

Suschak JJ, Williams JA, Schmaljohn CS (2017) Advancements in DNA vaccine vectors, non-mechanical delivery methods, and molecular adjuvants to increase immunogenicity. Hum Vaccin Immunother 13:2837–2848. https://doi.org/10.1080/21645515.2017.1330236

Syomin BV, Ilyin YV (2019) Virus-like particles as an instrument of vaccine production. Mol Biol 53:323–334. https://doi.org/10.1134/S0026893319030154

Tebas P, Yang SP, Boyer JD et al (2021) Safety and immunogenicity of INO-4800 DNA vaccine against SARS-CoV-2: A preliminary report of an open-label, phase 1 clinical trial. EclinicalMedicine 31:100689. https://doi.org/10.1016/j.eclinm.2020.100689

Tian JH, Patel N, Haupt R et al (2021) SARS-CoV-2 spike glycoprotein vaccine candidate NVX-CoV2373 immunogenicity in baboons and protection in mice. Nat Commun 12:372. https://doi.org/10.1038/s41467-020-20653-8

Tseng CT, Sbrana E, Iwata-Yoshikawa N et al (2012) Immunization with SARS coronavirus vaccines leads to pulmonary immunopathology on challenge with the SARS virus. PLoS One 7:e35421. https://doi.org/10.1371/journal.pone.0035421

vfa (2021) Vaccines to protect against COVID-19, the new coronavirus infection. https://www.vfa.de/de/englische-inhalte/vaccines-to-protect-against-covid-19. Accessed 18 März 2021

Vogel PUB, Schaub GA (2021) Seuchen, alte und neue Gefahren – Von der Pest bis COVID-19. Springer Spektrum, Wiesbaden. https://doi.org/10.1007/978-3-658-32953-2

Volkers P, Poley-Ochmann S, Nübling M (2005) Regulatorische Aspekte klinischer Prüfungen unter besonderer Berücksichtigung biologischer Arzneimittel. Bundesgesundheitsbl 48:408–414. https://doi.org/10.1007/s00103-005-1014-9

Voysey M, Clemens SAC, Madhi SA et al (2021a) Safety and efficacy of the ChAdOx1 nCoV-19 vaccine (AZD1222) against SARS-CoV-2: an interim analysis of four randomised controlled trials in Brazil, South Africa, and the UK. Lancet 397:99–111. https://doi.org/10.1016/S0140-6736(20)32661-1

Voysey M, Clemens SAC, Madhi SA et al (2021b) Single-dose administration and the influence of the timing of the booster dose on immunogenicity and efficacy of ChAdOx1 nCoV-19 (AZD1222) vaccine: a pooled analysis of four randomised trials. Lancet 19 S0140-6736(21):00432–00433. https://doi.org/10.1016/S0140-6736(21)00432-3

Vujadinovic M, Vellinga J (2018) Progress in adenoviral capsid-display vaccines. Biomedicines 6:81. https://doi.org/10.3390/biomedicines6030081

Wang F, Kream RM, Stefano GB (2020) Long-term respiratory and neurological sequelae of COVID-19. Med Sci Monit 26:e928996. https://doi.org/10.12659/MSM.928996

Weisblum Y, Schmidt F, Zhang F et al (2020) Escape from neutralizing antibodies by SARS-CoV-2 spike protein variants. Elife 9:e61312. https://doi.org/10.7554/eLife.61312

Wen J, Cheng Y, Ling R et al (2020) Antibody-dependent enhancement of coronavirus. Int J Infect Dis 100:483–489. https://doi.org/10.1016/j.ijid.2020.09.015

WHO (2004) Summary of probable SARS cases with onset of illness from 1 November 2002 to 31 July 2003. https://www.who.int/csr/sars/country/table2004_04_21/en/. Accessed 30 January 2021

WHO (2020a) Draft landscape of COVID-19 candidate vaccines. file:///C:/Users/2517833/Downloads/novel-coronavirus-landscape-covid-19.pdf. Accessed 3 February 2021

WHO (2020b) MERS situation update. https://www.emro.who.int/health-topics/mers-cov/mers-outbreaks.html. Accessed 20 February 2021

Wong AHM, Tomlinson ACA, Zhou D, Satkunarajah M, Chen K (2017) Receptor-binding loops in alphacoronavirus adaptation and evolution. Nat Commun 8:1735. https://doi.org/10.1038/s41467-017-01706-x

Wu LP, Wang NC, Chang YH, Tian XY, Na DY et al (2007) Duration of antibody responses after severe acute respiratory syndrome. Emerg Infect Dis 13:1562–1564. https://doi.org/10.3201/eid1310.070576

Ye ZW, Yuan S, Yuen KS, Fung SY, Chang CP et al (2020) Zoonotic origins of human coronaviruses. Int J Biol Sci 16:1686–1697. https://doi.org/10.7150/ijbs.45472

Zaki AM, van Boheemen S, Bestebroer TM et al (2012) Isolation of a novel coronavirus from a man with pneumonia in Saudi Arabia. N Engl J Med 367:1814–1820. https://doi.org/10.1056/NEJMoa1211721

Zhao H, Shen D, Zhou H et al (2020) Guillain-Barré syndrome associated with SARS-CoV-2 infection: causality or coincidence? Lancet Neurol 19:383–384. https://doi.org/10.1016/S1474-4422(20)30109-5

Zündorf I, Dingermann T (2017) Vom Hühnerei zur Gentechnologie. Pharmazeutische Zeitung; https://www.pharmazeutische-zeitung.de/ausgabe-132017/vom-huehnerei-zur-gentechnologie/. Accessed 30 June 2020

GPSR Compliance
The European Union's (EU) General Product Safety Regulation (GPSR) is a set of rules that requires consumer products to be safe and our obligations to ensure this.

If you have any concerns about our products, you can contact us on

ProductSafety@springernature.com

In case Publisher is established outside the EU, the EU authorized representative is:

Springer Nature Customer Service Center GmbH
Europaplatz 3
69115 Heidelberg, Germany

www.ingramcontent.com/pod-product-compliance
Ingram Content Group UK Ltd.
Pitfield, Milton Keynes, MK11 3LW, UK
UKHW042058021026
15589UKWH00001B/49

* 9 7 8 3 6 5 8 3 8 9 3 2 1 *